LONDON'S PRE-WAR SMALLER CLASSES

LONDON'S PRE-WAR SMALLER CLASSES

C, STD, LTC, TF, CR

KEN BLACKER

Capital Transport

First published 2022

ISBN 978-1-85414-473-7

Published by Capital Transport Publishing Ltd
www.capitaltransport.com

Printed in the EU

Cover The STDs were the first non-AEC vehicles ever to be purchased by London Transport for its double deck bus fleet. As a result, their introduction created a great amount of interest. and they were prolifically photographed in their early days. Many in London Transport's senior engineering management held out high hopes for them, but no-one could have predicted how extraordinarily successful they would prove to be. STD 12 was one of the very first in service when Hendon's complete allocation on route 52 was taken over by the new Leylands on 23rd and 24th April 1937 and it was photographed when still brand new. With the exception of two spells at Cricklewood garage during the early war years, STD 12's entire working life was spent at Hendon, where it was finally made redundant on 1st December 1953. Like all of its class, STD 12 was almost as reliable to operate and structurally as strong at the end as it had been when it was new more than 16 years earlier. *London Transport Museum*

Title page The sturdy and characterful Leyland Cubs proved to be a wise choice for one-man operation and they formed the backbone of the 20-seater fleet from 1935 to 1953. A typical member of the class, C 16 was photographed in Market Street, Dartford, outside the Ministry of Labour and National Service, a reminder of the wartime years which continued in being until as late as 1959 when compulsory military service was finally abolished. C 16 had led an interesting career, having been sent out on loan in 1942/3 and changing livery twice; from green to red in April 1942 and back again in 1947. Its only settled spell of operation whilst with London Transport was at Northfleet from April 1946 to May 1953, and it subsequently worked in Ceylon from January 1955 through to June 1964, reaching a very creditable age of 29 when its retirement finally came about. *Capital Transport*

CONTENTS

The distinctive and characterful TFs retained their reputation for speed and reliability right to the end, even though spare parts to keep them going were sometimes hard to find in their later days. Approaching 14 years old and demoted to bus work in March 1952, TF 44 ended its career at Hertford until its ultimate withdrawal on 1st June 1953. It was one of many TFs which, despite only having a few months of working life ahead of them, received a fresh coat of paint in their final year. *Mick Webber collection*

INTRODUCTION

Wind-down gas lamps, a typical nineteen-thirties feature on main roads in many parts of London, proliferate on Watford Way in the vicinity of Apex Corner where newly licensed STD 37 heads towards its terminus at Edgware station. The LEYLAND name plate at the base of the radiator lasted for only a very short time before being removed from all vehicles, perhaps in deference to AEC whose monogram was never displayed on the front of London Transport buses. The 'via' blind is a temporary substitute for the correct one which presumably has still to be supplied by the printing shop at Chiswick. *J M Aldridge collection*

Getting old has many disadvantages, but there is a big plus side to it too in the form of a host of treasured memories. Memories of people, places and – to someone who has been bus and tram mad like myself since a very early age – a host of wonderful vehicles that I followed, rode on, photographed and documented with a zeal that barely diminished as the years went by. These vehicles formed an unremarkable but dependable part of everyday life in those simpler, pre-computer-age days, but like everything and everyone, they all eventually reached the end of the road and moved on. Fascination with them led me to a career working for London Transport, a massive organisation which, despite some glaring shortcomings that tested the public's patience from time to time, remained arguably one of the greatest co-ordinated public transport networks that ever existed.

I first started recording London Transport's road vehicles in detail in this Capital Transport series way back in 1979, starting with the RT family, and this volume covering the many smaller classes from the period 1933 to 1939 and not so far dealt with, now almost completes my task. The post-war smaller classes will be covered in the last of these books, preparation of which is in progress.

Who is likely to be interested in vehicles which disappeared from the streets half a century and more ago? I don't know the answer to this, but I have a firm belief that all aspects of history should be documented as fully as possible before it is too late, and this includes everyday things as mundane and prosaic as bus fleets. I sincerely hope that there will be people out there who will find that the subject matter of this volume interests and stimulates them just as much as it has me.

In the realm of bus development, London Transport's rolling stock department was famous right from the Board's inception in 1933 for two things. One was its quest to be at the forefront of bus design, which it undoubtedly was for its first two decades, and the other was its ever-expanding search for standardisation, resulting in massive fleets of look-alikes in the form of the RT family, the Routemasters and, in single deck form, the RFs. In this volume the extremes of its obsession to be at the cutting-edge of vehicle development are recorded through the incredible success of its TF class set against the abject failure of the CRs. High levels of standardisation, which were only hinted at with the huge pre-war STL fleet but neared fulfilment with the RTs, offered great financial and other benefits, but they could also lead to built-in obsolescence and an inability to react to sudden market changes, as witnessed when the post-war swings to lightweight vehicles and one-man operation came fully into effect. Inevitably, despite its preference for huge, look-alike fleets, London Transport was obliged by factors beyond its control to purchase smaller classes of vehicle from time to time.

The information contained in this volume has been amassed over many decades from a whole variety of documented sources, official and unofficial, and it also includes memories gleaned from a host of friends, work colleagues and others over the years, very many of whom are now no longer with us. It has, of course, helped that – with very few exceptions – I knew and travelled on all the different types of vehicle described herein, so I knew their quirks, their plus points and their downsides, the memories of which have never left me. I have endeavoured to make this as complete a record of each individual class as possible, but inevitably there will be gaps in the accumulated knowledge and there may well be odd inaccuracies either in interpretation or even in the source material itself from which information is taken for which, of course, I apologise but have no control over. Just one word of caution. Throughout the text I have included route details where appropriate, but these are meant only as a guide and are not intended to be complete records of all services on which the vehicles operated or of the dates on which they operated them. You will need to look to other sources for this information.

As ever, I wish to express my grateful thanks to everyone who has helped me over the years by supplying information, memories or photographs which have been invaluable to me in completing this volume. Special thanks go once again to Alan Nightingale who has made great efforts to unearth photographs and additional information and who kindly took on the task of checking the draft of the text for the inevitable grammatical errors and typos. Friends from the past who have contributed posthumously are the late John Aldridge, Peter Aldridge, Frank Davis, John Gillham, Ken Glazier, Les Hampton, Don Jones, Vic Jones, Ron Lunn, Prince Marshall, George Robbins, Allen T Smith, Don Thompson and Reg Westgate. Those still with us include Laurie Akehurst, Tony Beard, Alan Cross, Dave Jones, Hugh Taylor, Mick Webber and Jim Whiting. A big 'Thank You' to you all.

KEN BLACKER

A day out at Chessington Zoo with London Transport's tours department, summer 1939 style, could include a trip in one of the very latest and most modern of sightseeing coaches which, in this instance, was TF 2. War was soon to bring such pleasures to an abrupt end. *Andrew N Porter*

THE C CLASS

ONE-MAN BUSES C 1 – C 98

For almost two decades the busy little Leyland Cubs represented the face of London Transport on services where low capacity, one-man operated buses had to be used, either because road conditions prohibited the use of larger vehicles or because passenger demand warranted nothing bigger. They negotiated twisty, narrow country lanes and climbed hills with apparent ease, and their characteristic engine roar seemed to signify a level of power and dependability out of proportion to their diminutive size. They provided a lifeline to many small communities in days before the omnipotence of the private car. The active lifespan of the C class, which effectively covered the years between 1935 and 1953, marked them out as being one of London Transport's more notable achievements bearing in mind that, when they were new, a six-year life was all that was predicted for them.

The genesis of the C class can be dated right back to the formation of the London Passenger Transport Board on 1st July 1933 and the months immediately following this when the Board's engineers, who were effectively the former London General Omnibus Company's staff working under a different name, were forced to confront the future of the bus and coach fleet which was destined to grow much larger with the forthcoming acquisition of vehicles from a multiplicity of private operators under the compulsory purchase provisions provided by the London Passenger Transport Act. It was known in advance that many of these acquired vehicles would fall into the category of 20-seaters or smaller, representing a multitude of different chassis makes, ranging in physical condition between lovingly well-maintained right down to truly decrepit, depending upon the professionalism and financial standing (or lack of each) of their owners. It was necessary

Photographs taken of C 1 before entering service show the developing thought processes at Chiswick on minor aspects of the design. The earlier photograph, taken on the test gradient in Chiswick Works generally referred to as 'The Dip', shows that as originally built it had no front bumper. When the later photograph (opposite) was taken just after the vehicle was taxed and ready for road trials a bumper had been fitted with the registration plate hanging below it. A most unusual feature, seen clearly in the earlier shot, is the strange positioning of the winding handle for the destination blind projecting out of the side of the vehicle just in front of the saloon door. The later scene shows it repositioned, still externally but now adjacent to the destination box itself but with no provision made for the driver to be able to reach it conveniently. It had been returned to position number one before C 1 finally entered service.
Tony Beard collection / London Transport Museum

to plan for their absorption into the main London Transport fleet and then for their replacement by a standardised stock of new 20-seaters in line with the policy of renewing the entire fleet over time with vehicles of standardised types.

London Transport did not begin life in July 1933 with a very big fleet of 20-seaters. The Central Bus department inherited 53 from the LGOC (comprising 42 of its tried and trusted Dennis Darts along with 7 Guys and 4 Beans). The Country Bus & Coach department began with a surprisingly small contingent – a mere 36 – which all came from London General Country Services and were a mixed bag of ADC (2), Bean (2), Chevrolet (1), Commer (6), GMC (2), Guy (8), Morris (10) and Thornycroft (5), of which only the five Thornycroft Nippys really impressed their new owners. However, things were quickly to change.

To increase the number of one-man buses available for service twelve Gilford 26-seaters, which originated from Skylark, were downseated from 26 to 20. Then, starting with the first acquisitions on 1st November 1933 through to the end of 1934, a substantial fleet of small capacity, one-man vehicles built up, representing a staggering variety of different makes. By the end of 1934 a total of 290 small buses and coaches had accumulated, and in some cases had been immediately delicensed because of their run-down condition or because of the near-impossibility of acquiring spare parts for them at a reasonable price. The following list shows no fewer than 24 different makes of vehicle as joining the fleet over the 18-month period, all with seating capacities varying between 14 and 20.

ADC	2	GMC	9
AJS	1	Guy	18
BAT	1	International	1
Bean	12	Leyland	1
Bedford	25	Manchester	6
Brockway	1	Morris	18
Chevrolet	32	Overland	1
Commer	13	Reo	5
Crossley	4	Saurer	2
Dennis	85	Star	3
Dodge	7	Thornycroft	14
Ford	17		
Gilford	12		

During the autumn of 1933 the Board's engineers took a preliminary look at their possible future vehicle requirements. With a goal of achieving some sort of standardisation of vehicle types within six years, they set out a programme of possible new vehicle purchases to cover an initial two-year period between 1st July 1934 and 30th June 1936. This would allow time for the first orders to be placed and built. As far as 20-seaters were concerned the initial plan called for the purchase of 12 Thornycroft Nippys, 12 Dennis Darts and 50 Leyland Cubs. The engineering team had also taken into consideration the Bedford WLB but decided against it in view of its lightweight construction. However, the Board was subsequently content to use acquired vehicles of this make, not only as lorries and general service vehicles but also as buses, a role which some performed until the outbreak of war.

The reason for favouring the Thornycrofts and Dennises was clear: these were known to be good, reliable vehicles. However, the reason for recommending Leyland Cubs for the bulk of the order was less obvious, even to a doubting Frank Pick who instructed that a vehicle of this type should be thoroughly examined and tested before any such order was placed. The Board had, in fact, inherited a single Leyland Cub, JH 2401 (which subsequently became C 76) when it took over the St Albans & District business on 1st November 1933, but it is not known what, if any, influence this had on the decision. A much more likely explanation is that, having made contact with Leyland and its development engineers, LT's representatives were impressed with what they had seen and heard, not only in relation to the Cub model but also with regard to Leyland's future development plans, which they found to be of great interest.

Although London Transport was tied by its contractual obligation to purchase the bulk of its new motor buses from AEC, it was free to go elsewhere for a minority of purchases if good reason could be found for doing so. Also, with the prospect of having to purchase large numbers of new trolleybuses looming on the horizon, and with no pre-existing contractual restrictions applying with this type of vehicle, Leyland was viewed as being one of the few suppliers with sufficient capacity to meet the Board's need for such vehicles in years to come so it was necessary to maintain strong relationships between the two organisations.

Leyland willingly agreed to let London Transport borrow a Cub chassis for examination, and KP3/2257 duly arrived at Chiswick on 15th January 1934 where intensive tests were carried out. It was unbodied at the time, but after returning to Leyland's after a stay of almost five months it was bodied by Willowbrook and subsequently entered the Skill's fleet in Nottingham registered AAU 947.

The Cub was a fairly new model, having gone into production in 1931 after construction of a few prototypes in the previous year. It was assembled at Leyland's plant near Kingston-upon-Thames which had been totally and expensively re-equipped in 1931 to produce the Cub range. Kingston was an assembly rather than a manufacturing plant, the Cubs consisting totally of brought-in parts. Some of these, such as the engine and axles, came from the Company's main works in Lancashire, while others, such as the chassis frame and steering gear, were sourced from outside manufacturers. The design was simple but sturdy, with a constant mesh gearbox including 'silent' third gear with the gear-change lever on the driver's right-hand side, single plate clutch, and Lockheed four-wheeled braking without servo assistance. The 27.3hp 6-cylinder petrol engine was of the now-dated side valve variety with the crankcase and cylinders being integral and of unit construction with the clutch and gearbox. Bore and stroke were 3⅜ins and 5ins respectively, and the RAC rated 27.3hp engine ran smoothly and was well capable of handling a 20-passenger load in the 15ft 6ins wheelbase KP3 chassis London Transport was interested in.

Without waiting for any test results to be obtained, a Special Expenditure Requisition was approved on the very same day that the test chassis arrived at Chiswick, 15th January 1934, for a single trial vehicle to be purchased, for which design work commenced almost immediately on a composite framed body that London Transport itself intended to build. A sum of exactly £1,000 was budgeted for the complete vehicle. As things subsequently turned out, this was destined to be the only Cub body ever manufactured at Chiswick.

The new chassis (no. KP3/2954) was built to comply with strict specifications laid down by London Transport which Leyland was happy to go along with, especially as a heavy hint had been dropped that, if this vehicle proved successful, an order for up to 100 could follow of which 60 would have petrol engines and 40 would be oil powered. Leyland had recently developed an oil engine for the Cub and London Transport was well aware of this even though it specified petrol for this initial purchase. The new oil engine was a completely new design with no physical resemblance to the side valve petrol unit used in the Cub except that both shared the same cylinder bore and stroke dimensions of 3⅜ins x 5ins, retaining the same 4.4 litre capacity. It was a direct injection pushrod-operated overhead valve unit which ran smoothly and quietly and had gone into regular production in September 1933. Prior to this a pre-production unit in a Cub goods chassis had achieved a creditable 25 miles per gallon on long runs.

The type classification C (standing for Cub) was London Transport's obvious choice for the new class, and the new chassis for C 1 arrived at Chiswick from Kingston on 11th June 1934. The body shops had by then almost completed the new body (no. 14756) which was mounted later in the same month. A very extensive period of testing was then undertaken before the vehicle was taxed as AYV 717 on 10th October 1934.

The Chiswick design department had produced a workmanlike body for C 1 which, with the bonnet blending neatly into the front dash and a well-rounded rear end, was thoroughly modern in appearance. The KP3 chassis had a wheelbase of 15ft 6ins, and within an overall length of 23ft 11½ins the vehicle was capable of accommodating 20 passengers plus driver comfortably with seats placed at a distance of 2ft 6½ins apart. The seating layout was quite complex. The front seat on the nearside, just behind the entrance door, was for a single person, behind which were two pairs

Viewed from the back while in Chiswick Works, C 1 exhibited a modern and well-rounded, if somewhat bulbous, rear end. As at the front, the framework to the destination box was picked out in white in stark contrast to the black roof panels. The double row of black rubbing strakes at the back – an anti-cyclist measure – was copied from current double deck practice on STLs but was not repeated on any further single deckers. *Tony Beard collection*

of conventional twin seats. On the offside there were three pairs of such seats. Twin seats faced inwards over the rear wheel arches, and there was a crosswise one seating five at the rear. The passenger door slid rearwards into an enclosed area and was operated by electrically controlled G D Peters vacuum door gear, the tank for which was carried on the side of the chassis frame. The emergency door was situated in the rearmost bay on the offside. Half-drop windows were provided by Hallam, Sleigh & Cheston and there were "Ashanco" ventilators in the roof. Unladen weight for the complete vehicle was 3tons 9cwt giving a laden weight (including a full passenger load and 30 gallons of petrol) of 4tons 17cwt, which was well within a maximum of five tons decreed by London Transport as being acceptable for 20-seaters on account of certain bridge restrictions. The total cost of purchasing C 1 had been £978 (£588 for the body and £470 for the chassis), which marked a considerable saving over the £1,030 originally budgeted for the project.

C 1 was painted in red livery which, as on Q 1, was enhanced by a black roof. It was initially allocated to Hounslow garage for operation on route 237 between there and Chertsey, and it immediately proved popular with drivers in comparison with the Dennis Darts that it ran alongside. Management, too, was satisfied because petrol consumption was 1.17mpg better than the earlier Darts and 0.7mpg improved over the later model. A small amount of carburettor trouble was experienced in the early days while, on the bodywork, several door problems were encountered and rainwater entered the cab area through a gap in the blind box. The front wings were initially made of aluminium which tended to split and they proved unable to withstand light pressures without denting. However, all these defects were soon rectified, leading to the official conclusion that the Leyland Cub was "highly satisfactory and far superior to the 'one man' buses now being operated".

C 1 is seen running from Hounslow during its early service days while still petrol engined. The location is Sunbury which, like much of rural Middlesex, suffered heavy development during the nineteen-twenties and thirties. Today the ornate clock tower still remains, but the environment around it has changed beyond all recognition and this air of bucolic tranquility has now been lost for good. C 1 stayed at Hounslow for about eight months until it went away for an oil engine to be fitted. *Mick Webber collection*

London Transport did not wait for the trials on C 1 to be concluded before making its final decision on future purchasing policy. Results of the early trials at Chiswick had already been sufficiently good to convince it that the future lay with the Leyland Cub to the exclusion of the Thornycrofts and Dennises included in the earlier plans. Dennis had, in any case, indicated that production of the Dart had ceased, and it was now marketing the Ace as its answer to the Leyland Cub. London Transport had, in fact, taken delivery of two new Aces in March 1934, AKR 937 and BPF 318 (subsequently numbered DC 1 and 2). These were a pair of Waveney-bodied 20-seaters which had been ordered by Gravesend & District before it had been taken over and were suitably modified during construction to meet the Board's requirements. Although the Ace proved popular with many other operators throughout the nineteen-thirties as a strong competitor to the Cub, London Transport showed no further interest in it. The idea of buying more Thornycrofts was pushed aside by the conclusion that the Cub was of superior design, though the final decision was probably clinched by the fact that an oil engine was now available in the Cub, and this was what London Transport really wanted.

In August 1934 an order was placed with Leyland for the supply of 74 Cub KPO3 oil-engined chassis (the order was actually for 75 vehicles as it covered the initial chassis C 1 as well as 74 new vehicles C 2-75). All-metal bodywork was to be built at Rochester by Short Brothers Ltd. A month later Leyland agreed to replace the petrol engine in C 1 with a similar oil engine to those intended for C 2-75 at an overall cost of £126 which would include alterations to the bodywork to incorporate the necessary 24-volt electrical system in place of the existing 12-volt arrangement. For reasons recorded later, this arrangement was never implemented.

This is an appropriate moment to review C 76, the oldest Cub in the London Transport fleet and the only one to be obtained second hand. It was the second from newest vehicle in the 14-strong St Albans & District Motor Services fleet owned by Charles Russett & Son which operated out of two garages in the town and was taken over by London Transport in an early wave of acquisitions on 10th November 1933. An early KP model, with chassis numbered 164, it had been registered from new as JH 2401 on 14th June 1932, but in common with most of the Country Bus & Coach department's fleet, it ran unnumbered until February 1935 when control of the vehicles within it passed from Reigate to Chiswick, hence the high fleet number. The St Albans & District operation had been a hybrid one, with some services running into the Metropolitan Police District and requiring buses working therein to carry Metropolitan Stage Carriage plates and to comply fully with the strict standards laid down by the Public Carriage Office. The Cub was not one of these, and it was used for tours, school contracts and excursion work and also as a replacement for service buses on routes which did not enter the Metropolitan Police District.

No photos are known to exist of C 76 during its St Albans & District days. The earliest shows it operating for London Transport from Slough garage carrying the original Country Bus colours of green and black. This photograph shows the vehicle in March 1935 after running number plates have been fitted but before a fleet number has been allocated to it. *J F Higham © A B Cross*

The manufacturer of C 76's 20-seat body is not known for certain but is thought to have been Thurgood's of Ware. It was of the usual composite construction. The seating layout consisted of three forward-facing pairs on the offside and two on the nearside ahead of inward-facing seats for three over the wheel arches. Double seats appeared again at the rear on either side of a central emergency door located in the back wall of the bus. Its initial livery was St Albans & District's light green and cream with gold leaf lettering, but soon after passing into London Transport ownership it was repainted into the Country area's standard much darker green and black with silver roof, probably initially carrying GENERAL fleet names. The vehicle was powered by the same type of side valve petrol engine as C 1 which it retained throughout its operating lifespan with London Transport.

When the type classification scheme was extended to Country area stock in 1935 C 76 became the sole 1/1C1/1, wrongly hinting at a similarity to C 1 (which was classified 1C1). In fact the two were very unalike. They looked entirely different, and whereas C 76's overall length was almost 1ft greater than C 1 (at 24ft 4ins) its wheelbase was shorter by 2 inches at 15ft 4ins. At 3tons 14cwt, its unladen weight exceeded that of C 1 by a full quarter of a ton.

Photographed after its first overhaul in September 1935, C 76 now carries Windsor garage plates and the later Country Bus colour scheme. The strengthening straps applied to the body pillars during overhaul are quite prominent and the rain shield above the windows now consists of a single narrow metal strip. The single step up to the saloon floor, and the screen behind the driver which was probably there from new, can be clearly seen through the open doorway. *A N Porter*

The earlier of these two offside views of C 76, wearing the later livery and standing over the pit in Windsor garage, probably dates from soon after its second and last overhaul in May 1937. The later view shows it carrying a newer and deeper type of Leyland Cub radiator with starting handle projecting through it. This photo was taken near the end of C 76's career in 1938 when it petrol engine was earmarked for putting into C 1. *D W K Jones / J F Higham © A B Cross*

Records have not survived to confirm exactly C 76's movements within the Country Bus & Coach fleet in its first two years but photographic evidence confirms that it operated out of Slough (London Road) garage at one time. On 31st July 1935 the vehicle went for its first full overhaul from which it emerged in the latest two-tone green livery, again with silver roof. It was off the road until 21st September during which time the body framework was extensively rebuilt and the side pillars reinforced with steel bracing strips to provide added strength. At the same time the glass window louvres were replaced with continuous metal strips and the side lights were moved from the front mudguards to a higher position on the front dash. On its return to service C 76 was allocated initially to Windsor, but later in 1935 was transferred to Harefield. This garage closed at the end of the year and C 76 passed to Watford High Street on 1st January 1936 before returning to Windsor on 10th December of the same year where it stayed for the remainder of its days with London Transport.

One further, brief body overhaul took place in May 1937, but C 76's non-standard nature meant that its days in the London Transport fleet were now numbered. On 28th October 1938 it was delicensed into Chiswick and on 31st January 1939 was sold to R A Jordan, a Biggleswade dealer who purchased many redundant London Transport buses in pre-war years. Early in December 1938, while it was at Chiswick awaiting its fate, a Perkins 4-cylinder oil engine was installed in it. This engine had previously been fitted experimentally to C 1 (as will be recounted later) and was no longer required, having perhaps broken down beyond repair. London Transport wished to re-convert C 1 to petrol propulsion for a further period of service and C 76's engine, being identical to the one that was in it when new, was ideal for this purpose. No record has ever come to light of C 76 ever actually running with the Perkins engine installed, or indeed of it being used again in any capacity after leaving London Transport, and it seems highly likely that Jordan's scrapped it.

After receiving the order for 74 more Cubs, Leyland did not waste much time in getting construction of a preliminary chassis under way. Completed by November 1934, this was chassis KPO3/3773 which was destined in due course to become C 2 in the London Transport fleet. It was quickly despatched to Chiswick where it underwent a series of tests carrying weights equivalent to a fully loaded bus with a full tank of fuel. Resulting from these trials, London Transport issued various specification amendments of its own for the production batch, the most notable of which was the fitting of a heavier flywheel to meet the Board's idling standards. The main production batch began going to the coachbuilder in early 1935 and carried chassis numbers KPO3/3960-4032, although these were not allocated to C 3-74 in numerical sequence.

The whole batch of 74 Cs was intended for Country area use and when the first of them were delivered from Short Brothers' factory in April 1935 they were some of the very earliest to carry the new Country Bus livery consisting of two shades of cheerful mid and light green separated by black moulding strips, replacing the rather sombre combination of dark green and black with silver roof. This was, in fact, the only batch of bus bodies ever built by Short's for London Transport, although its predecessors had bought from the Rochester-based company from time to time. As a result of its aircraft-building experience, Short's had gained considerable expertise in the use of light-weight metals for construction purposes and had gained renown as one of the early pioneers in the use of metal framework for road transport vehicles.

From the frontal aspect the 74 Cubs in the first production batch were impressive looking machines whose performance lived up to their strong, workmanlike looks. Front bumpers were fitted from new and the registration plate was situated above Leyland's 'Cub' insignia which remained in place even though the Leyland name was not itself carried on the radiator. C 58 was new to Windsor garage on 2nd July 1935 and remained there until February 1939 when it moved to Dunton Green, which became its permanent home until withdrawal in November 1953. *A N Porter*

Yet to be licensed, and with its whole 18 year working life still ahead of it, C 3 undergoes a comprehensive series of tests both unladen and also weighted down with the equivalent of a full passenger load, to assess its operational capability. Viewed from the nearside the Short Brothers body bears a strong resemblance to C 1 except that the windows now all have rounded corners, and the winding gear for the front destination box is no longer visible as it has been sensibly re-housed inside the vehicle. *A B Cross collection*

The first of the batch, C 2, photographed prior to delivery from the coachbuilder. A clear difference from C 1 is the repositioning of the emergency door to a forward position in the first passenger bay, while for the main production run that followed on from C 2 the lower diving strip was added to the emergency door itself as well as to all the other lower panels. The outward flare at the back with the inbuilt bumper below it was to become a trade mark for pre-war London single deckers. *Alan Nightingale collection*

In their general shape C 2-75 fairly closely resembled the prototype C 1 although they had a more pronounced and modernising outward curve at their rear bottom edge which probably accounted for their slightly longer overall length of 24ft 0ins based on the same 15ft 6ins wheelbase. Although the difference was barely noticeable, at 8ft 10½in in height, they were in fact 1½ins taller than C 1. The main clearly observable difference, apart from a longer and more modern Cub radiator, was that the emergency exit was now situated immediately behind the driver instead of being located in the last offside bay.

The use of metal framework was accompanied by windows mounted in light steel pans which were radiused on all corners, another neat and modernising feature which distinguished the new buses from C 1 when viewed externally.

The most startling difference lay inside the saloon, where the usual proliferation of wooden mouldings forming the cant rail, pillar and waist rail mouldings had been replaced by modernistic, pressed aluminium cappings suitably rounded to suit the window profile and linked by small, neat 'bracelets', some of which were unpainted and polished where they separated different coloured rexine coverings. This was destined to be the standard style of interior finish for all single deckers built to London Transport specification for the next two decades although it did not come into regular use on double deckers until the RT class made its debut. The now-standard brown and pale green combination of internal decor featured on C 2-75, and the ceilings were in cream enamel picked out with green rexine-covered mouldings. The seats were upholstered in the same green moquette as currently used on new STLs and were mostly of the same wooden construction with brown and green painted backs and detachable aluminium top grab rails where relevant. A strange exception was the front nearside single seat which did not match the others and was built to the earlier 'bucket' style once used on LTs, STs and Dennis Darts. Unusually, there were no internal lining panels below the windows, which had probably been omitted to enable the vehicles to be kept down to an unladen weight of 3tons 19cwt despite the extra heaviness of the chassis caused by the use of an oil engine.

The interior of the Cub body looked quite spacious for a vehicle with only 20 seats. The view looking forward, taken on an unknown vehicle when new, shows clearly how the single front seat was of an earlier 'bucket' style than all the others. The view facing rearwards is of C 40 and was taken in January 1953 by which time all remaining Cubs carried standard post-war bus-style upholstery. The rear end arrangement included single, inward-facing seats over the wheel arches and a bench for five at the rear. *London Transport Museum / John C Gillham © Tony Peters*

Classified as 2C2, the new Cubs were allocated body numbers 14916-14989 and a batch of 74 registration numbers BXD 627-700 was reserved for them with the London County Council. The first six were numbered out of order (C 2-7 – BXD 631/627-629/632/630), but the remainder were in strict numerical sequence from BXD 633 upwards. The very first ones were licensed for service on 17th April 1935 and were spread between three garages: Addlestone, Northfleet and Windsor. Subsequent garage introductions were:

2nd May	Dunton Green
13th May	Epping, Hertford and St Albans
28th May	Staines
3rd June	Chelsham
7th June	Guildford
14th June	Dorking, Leatherhead
2nd September	Amersham

The last vehicle in the batch to enter service was C 71 at Hertford on 4th September 1935, and this brought the final allocation of Cubs to garages as follows:

Addlestone	12	Hertford	13
Amersham	3	Leatherhead	1
Chelsham	1	Northfleet	9
Dorking	2	St Albans	11
Dunton Green	3	Staines	4
Epping	1	Windsor	8
Guildford	6		

Seen in a typical country setting, in this case in the far south-east of the Country Bus area, is C 69 which was one of nine Cubs licensed new at Northfleet in two batches, three in April and May 1936 and six in July and August. C 69 entered service on 3rd August, but it ran only from the one-time tram depot in Dover Road East for a short while and had moved on to Dunton Green before the fine new Northfleet garage was opened in July 1937. *Mick Webber collection*

As intended, Cubs made heavy inroads into the miscellany of one-man small vehicles that London Transport had retained in service, all of which had received fleet livery but in most cases had remained without fleet numbers. However, the number of new Cubs fell well short of full scheduled requirements, especially on Saturdays when enhanced schedules were run on many services, so a number of older vehicles still remained in stock. These consisted principally of BD class Bedfords, which could be found scattered throughout the Country fleet, along with a number of NY class Thornycroft Nippys which were eventually concentrated at Addlestone and MS class Morrises which formed the bulk of the 20-seater fleet at Amersham. The two modern Dennis Aces also remained in service, and when a year-old Dennis Mace (BBH 755) was inherited from the Penn Bus Company on 1st August 1935 this, too, was kept and numbered DC 3.

The Cubs proved instantly popular with the travelling public but were not quite so well received by the driving staff who found them to be somewhat sluggish and less responsive than many of the vehicles they had replaced. London Transport took the matter seriously and informed Leyland that the Cubs appeared to be underpowered. Leyland's engineering department put the blame squarely on London Transport who, mindful of the hugely adverse public reaction that earlier diesel powered buses had provoked on account of their acrid black exhaust smoke, had gone to the opposite extreme and insisted that the Cubs should be totally smoke free. To achieve this, their fuel injection pumps had been governed to levels which adversely affected their power output. The Board's engineers ultimately solved the problem by adjusting the pump settings but petulantly insisted that, although these now slightly smoky vehicles might be acceptable on Country services, they would not be suitable for operation within the Central Bus fleet.

As it happened, Leyland had been working since about the middle of 1934 on a more powerful engine for the Cub at the request of a number of customers. They were looking at increasing the bore size to 3½ins, which was about as far as they could go within the constraints of the existing design of their Light Six engine, a modification which would increase the swept volume from 4.4 to 4.7 litres. London Transport was aware of this but was also intrigued by a new oil engine manufactured in Peterborough by F Perkins Ltd, a small and comparatively new manufacturer which had only been in business since 1932 but was already building up a strong reputation for the quality of its engines and had clinched a deal to fit them in Commer vans. London Transport's engineers had studied a Perkins engine in a Hillman staff car based at Chiswick and were highly impressed by its performance and reliability, and they now wanted to try one out in a Leyland Cub. C 1 was selected for this purpose, and in May 1935 the Board's arrangement (mentioned earlier) with Leyland to exchange C 1's petrol engine for an oiler was cancelled. Instead, the vehicle was sent to Perkins in June 1935 for its new Perkins 'Leopard' engine to be installed at a maximum outlay of £240.

The 'Leopard' engine was a very compact unit with only four 100mm x 127 mm cylinders which were in a Monobloc casting with the crank case. The overhead valves were operated by a high crankshaft and the Perkins-patented combustion system incorporated a long parallel throat running tangentially into an air cell. The engine required only a 12-volt starter, thereby avoiding the major rewiring usually encountered when converting from petrol to oil, and the dynamo was, in fact, removed from C 1's petrol engine for use on the new Perkins unit. At Perkins' suggestion the engine was flexibly mounted using the same system as employed on the Dennis Ace, and at the rear the gearbox rested on a serrated rubber pad to further insulate the power unit from the frame. Interestingly, London Transport kept Leyland fully informed of its intention to trial the Perkins engine and even offered them the chance to bench-test it, but it is not known if Leyland took up the offer.

Coincidentally, at about the same time that C 1 was being despatched to Peterborough, London Transport received a call from Leyland asking if a Cub could be provided to which its prototype more powerful new oil engine could be mounted to gain testing and service experience. C 51 was taken straight out of service at Northfleet and sent to Leyland's works at Kingston for the new engine to be fitted. Designated by the manufacturer as type

Sporting a 'Perkins Diesel' badge prominently on its radiator grille, C 1 looks otherwise unaltered from its petrol-engined days when allocated to Barking garage between August and November 1935 for operation at Romford on local services 238 and 252. The front registration plate has now once again been hung below the bumper. *J F Higham © A B Cross*

H2, the new engine was based structurally on the earlier model but had a modified cylinder head and block to accommodate an air cell. With an indirect injection combustion system it gave a rated power of 65bhp with swept volume of 4720cc, some 18% greater than the Perkins unit at 3990cc. In its new guise C 51 was classified as 1/2C2.

C 1 and C 51 were tested extensively at Chiswick during July 1935. Although it was not practical to carry out fuel consumption tests, many other comparative trials were undertaken. Initially C 1 retained its rear axle ratio of 5.5:1, but to obtain better results this was subsequently amended to 6.5:1, after which a highly creditable performance was obtained for a small 4-cylinder engine. The vehicle was deemed to be perfectly suitable for service operation with a smoke free exhaust and a maximum road speed of 41mph. On C 51 a governor was fitted which limited the top speed to 36½mph. Pump troubles initially clouded its performance, but after rectification it was concluded that the H2 engine was an improvement on the direct injection model, with acceptably clean exhaust and noticeably greater power output. Directly comparing the Perkins engine with the Leyland, their acceleration was found to be roughly equal up to about 24 mph. Under load the flexible mounting of the Perkins engine eliminated high frequency vibration and was smoother than the Leyland. In an official report issued subsequent to the trials the Perkins engine was stated to have performed "extraordinarily well" despite a minor criticism that it idled rather fast, giving the impression of fussiness.

The test results proved, in fact, to be purely academic because by the time they had been completed a decision had already been taken to fit the Leyland H2 engine in the next 20 Cubs which were ordered in July 1935 at a cost of £1,127 each including bodies. Before compiling the specification for these, London Transport had enquired about the difference in capital cost between buying chassis complete with the latest Leyland oil engine, or alternatively minus engine to enable the Perkins unit to be specified. It quickly became apparent that the overall cost would be considerably inflated by using Perkins engines, while consideration also had to be given to the limited facilities which a small firm like Perkins might be able to offer in the event of an epidemic of trouble. However, this did not deter the engineers from continuing to carry out experiments on C 1, and it spent from 9th to 27th February 1937 with Perkins at Peterborough and CAV at Acton undergoing further development work.

This second batch of Cubs (C 77-96) was destined for Central Buses, where it was intended that they would replace 2 Beans, 5 Guys and 7 miscellaneous Dennises in addition to providing scope for future service enhancements. Short Brothers declined a request to build the bodies for these, having now abandoned bus body production following an upsurge in their traditional trade of aircraft manufacture, and as a result Weymann was contracted to build metal-framed bodies as nearly identical as possible to those on C 2-74. In November 1935 an order for two more was placed to provide an overhaul float, potentially bringing the fleet of one-man Cubs up to its peak level at C 98. At a total outlay of £1,070 per vehicle, C 76-98 had cost almost £100 more than C 2-74 which were purchased for £978, with both chassis and body costs having risen from £470 to £545 (chassis) and £508 to £525 (body). In addition to the revised engine, London Transport had specified certain chassis modification for the second batch, including gravity feed chassis lubrication and revisions to the clutch, all of which probably accounted for the rather sharp increase in chassis price.

It was now time to put C 1 and C 51 back into service to ascertain how efficiently their new engines operated in real life. C 1 resumed duties first, arriving at Merton garage on 3rd August 1935 and now carrying a Perkins badge prominently on its radiator. After only a few weeks it was transferred on 21st August to Barking with the intention of operating it alongside various Dennis Darts on Romford local service 252. C 1's service on the 252 was cut short on 9th November 1935 when it was transferred to Northfleet as a much belated replacement for C 51. In due course it was repainted from red to green, but it is not known exactly when.

Meanwhile problems had arisen with the engine on C 51 which meant that it was much slower returning to service than C 1. As C 51 was, in effect, the prototype for the forthcoming Central Bus fleet of Cubs, it was decided to operate it there right

After only three months of Country area service C 51 was re-engined and painted red in November 1935, duly replacing C 1 on Barking's allocation roster. With the exception of second-hand C 76, it was the only Cub ever to appear with a silver roof. It stands, accompanied by a couple of STs, in the spacious 1924-built surroundings of the former Romford garage which had been officially renamed Hornchurch on 1st September 1935, shortly before this photograph was taken. *London Transport Museum*

from the start, and its transfer to Central Buses meant that it had to be counter-balanced by the move of a similar type of vehicle in the opposite direction, hence C 1's reallocation to Northfleet. C 51 finally re-entered service, newly repainted red, on 28th November 1935, initially allocated 'on paper' to Barking and transferred to Hornchurch on 9th December 1936 when the latter assumed responsibility for the local one-man operations. Unlike C 1, its roof and rear dome were painted silver instead of gloss black, but this proved to be a one-off amongst Central Bus Cubs, and when C 77-98 were built they reverted to the black roofs.

Just before entering service on route 252, C 51 had returned from a re-call from Leyland's works – the latest of several – for modifications to be carried out, but teething troubles continued to afflict it and it is probable that it was during a further visit to Kingston between 29th January and 5th February 1936 that a modified engine, known as the H2B, was fitted. This was the same type of engine with which C 77-98 were equipped when they were put into service in April and May 1936.

Before describing the arrival of the Central area Cubs in the spring of 1936 it is appropriate to take a glance at what happened to the Cubs in the Country area in that year. After a quiet and fairly static spell during their initial few months of operation, the Country Cubs began moving around with a vengeance in 1936 starting on 1st January with an initial allocation of four to Watford High Street, a move indirectly connected with the closure of Harefield garage. By the end of the year the number of Cubs at Watford High Street had built up to eight, while the number at Chelsham and Leatherhead also increased from their original one apiece by a further three. On the other hand Amersham lost all three of its Cubs on 1st January to supply the bulk of Watford High Street's initial requirements, and although two others were received at Amersham

Delivered to Addlestone at the start of July 1935, C 54 remained there throughout 1936 but moved north to Amersham in October 1937. It was destined to be one of the unlucky members of its class which only saw a few years' service, being withdrawn at Amersham on 1st November 1944.
London Transport Museum

in April they departed again in June. Dorking lost both of its Cubs in April and Epping's lone example departed in July, while Reigate was the recipient of a single vehicle which lasted for a mere 18 days in June. All these moves were counter-balanced by transfers of Bedfords and other small vehicles still remaining in the Country Bus department's fleet. Additionally, during the year a number of Cubs moved to new operational premises thanks to London Transport's programme of building splendid new garages in several country towns. The new Staines garage opened on 3rd June (its fleet having been housed temporarily in Egham since 8th October 1935 while a new garage was built on the old site), while Addlestone opened on a new site on 24th June followed by St Albans on 26th August.

Delivery to the coachbuilders of the chassis for the new Central area Cubs began on 24th January 1936. As far as chassis and registration numbers were concerned the 20 chassis originally ordered and the two subsequent additions were all treated as one batch, being KPO3/4465-4486 and CLE 105-126 respectively. Body numbers for C 77-96 were 16131-16250 with C 97/98 following shortly afterwards as 16254/16255. As with C 51, the 22 new vehicles were classified 1/2C2, and they had the same general mechanical specification including a rear axle ratio of 6.5:1. The major point of difference from the earlier batch – apart, of course, from livery – was that the front bumper was now omitted, allowing the unladen weight to fall by 7cwt to 3tons 12cwt. The Weymann bodies were almost indistinguishable from the earlier Short Brothers batch except that the running boards on each side were reduced in size. On the nearside this eliminated the need for the doorway pillar to flare outwards at the bottom because, with a smaller running board fitted, there was no need for it to do so. An additional feature was that roof-mounted route stencil holders were now provided on both sides above the cant rail in the first passenger bay. The Central Bus colours of red with off white (officially grey) window frames and black roof and mudguards replicated the livery on C 1.

The first of the batch to be licensed for service were C 78/79 at Enfield and C 80/82 at Mortlake, all on 3rd April 1936. Next day Barking received its first one, followed by Hanwell on 16th April and Harrow Weald on 2nd May. With the final licensing of C 97/98 at Enfield on 19th May the allocation per garage stood at Barking 3 (plus C 51), Enfield 8, Hanwell 4, Harrow Weald 5 and Mortlake 2.

The first of the Weymann batch, C 77, was selected as the subject for the official photographer even though the route stencil holder in the roof was not yet complete and lacked its white background. Although presented as an Enfield garage vehicle its first allocation was, in fact, Hanwell. The use of a smaller running board enabled the outward curve at the foot of the first pillar behind the door to be eliminated on this batch of Cubs. *London Transport Museum*

Roof mounted route stencils were carried on both sides of the Central Bus Cubs, and the offside location is demonstrated by Enfield's C 81 when laying over at the Potters Bar terminus with ST 718 from the local garage standing behind. Like the earlier Short Brothers batch, the Weymann bodies were embellished with chromium plated windscreen surrounds when new but this feature lasted in most cases only until the first overhaul when they were painted in the main body colour. *D W K Jones*

In pre-war years Mortlake garage kept two Cs, one for service plus a spare, to cover the 1½ mile long route 207 linking the Railway Hotel at Barnes with the gold club house in Richmond Park. The operator who originally pioneered the service, Henry Turner, had found passenger demand low and gone broke, and London Transport suffered from equally sparse passenger loadings. It came as no surprise that the 207 was one of the first services to be withdrawn soon after the outbreak of war, on 16th September 1939. Passengers were noticeably absent from C 80, one of the Weymann batch, on its journey through Richmond Park. *London Transport Museum*

At Enfield garage the new vehicles ousted the last of the DS class 20-seaters based on the Dennis 2½-ton chassis while the Dennises withdrawn from Barking were smaller 30cwt type 14-seaters. The new deliveries to Hanwell saw the end of the BN class Bean 14-seaters, while two DA class Dennis Darts replaced by Cubs at Mortlake went temporarily to Harrow Weald to oust the last G class Guy ONDs ahead of the arrival there of new Cubs four weeks later. From now onwards the Central Bus one-man fleet consisted solely of C and DA class 20-seaters.

On 7th July 1937 two events of note happened in the Country area. One was the opening of a new garage to replace the old one at Northfleet, providing a fine new operating environment for the fleet based there including its nine Cubs. The other was the production for staff use of the first allocation book in the same format as the traditional Central Bus one showing all services and frequencies plus the garages and types of vehicle operating them. From this publication it is possible to ascertain that out of the 17 garages running one-man buses, five had no Cubs amongst their allocation and that few of those with a Cub allocation relied solely on these to meet their full service requirements.

	Maximum 20-seaters scheduled	No. of Cubs in stock
Addlestone (WY)	13	7
Amersham (MA)	10	-
Chelsham (CM)	3	4
Crawley (CY)	1	-
Dorking (DS)	1	1
Dunton Green (DG)	6	3
Epping (EP)	4	-
Grays (GY)	2	-
Guildford (GF)	5	6
Hertford (HG)	9	10
Leatherhead (LH)	4	3
Northfleet (NF)	8	8
Staines (ST)	3	2
St Albans (SA)	11	13
Watford High St (WA)	10	9
Watford Leavesden Rd (WT)	1	-
Windsor (WR)	11	9

Within a few months this situation was destined to change. On 6th October 1937 major service revisions at Addlestone saw a mass exodus of 20-seaters in favour of larger single deckers, leaving just one solitary Cub to cover its future requirements. Addlestone had been home to the last remaining NY class Thornycrofts, which were now withdrawn from service, while the majority of its Cubs were despatched to establish a presence at Amersham which had so far been the home of a small fleet of MS class Morris Viceroys. Opportunity was also taken to withdraw the newest Dennis 20-seaters in the fleet, the three-strong DC class consisting of two Aces and a forward control Mace.

Further changes were not far off and on 9th February 1938 T and Q types were drafted into St Albans to replace the vast majority of its 20-seaters, leaving just three Cubs there to service future requirements. Over the next few days the surplus Cubs drifted away. Two went to Crawley, which had never previously operated Cubs, and the remainder became part of the Amersham fleet, allowing the last of the Morris Viceroys to be withdrawn. Henceforth, the fleet of 20-seaters in the Country area consisted solely of Cubs helped out by a sprinkling of Bedfords.

The C class oilers were now performing extremely well on Country services, even if their exhaust emissions were a little hazy at times after the fuel pumps had been adjusted to provide a little more 'poke'. Unfortunately, the same could not be claimed for C 51, 77-98 in the Central Bus fleet. After what initially seemed a reasonably promising start, the indirect injection engines on these vehicles started to become troublesome in a variety of expensive ways including premature piston and bearing failures, and bottom end and even complete engine seizures.

London Transport was not alone in its troubles, which had also been reported by numerous other operators who were using the same engine both on passenger work and for goods haulage. Leyland's engineers worked hard to find solutions to oil flow and other troubles, but despite all their efforts no really lasting outcome was ever achieved and the fact remains that, in London service at least, the H2B engine never proved really satisfactory.

On a purely domestic basis, drivers soon began complaining about lack of power even though the new engine should theoretically have given a better performance than its predecessor. These complaints led to controlled tests being carried out at Chiswick on C 82 during August 1938, and the conclusion to this was a revision to the pump setting in what was effectively a re-run of what had occurred earlier on C 2-75.

In the Country area, C 1 came back into the spotlight in 1938. It had been running from St Albans garage since March 1937, latterly serving mainly as a back-up to the prototype rear-engined Cub CR 1 which was not proving as reliable in service as had originally been hoped. On 30th August C 1 was delicensed and subsequently taken into Chiswick Works for the future of its experimental Perkins engine to be decided. As recorded earlier in this chapter, the engine was taken from it and placed into C 76 which was now withdrawn from service and about to be sold. C 76's petrol engine, which was exactly the same type of side valve unit as had originally powered C 1 when new, was earmarked for fitting into C 1 to give this vehicle a renewed lease of life. An alternative to this, which would be to fit a Leyland diesel engine as used in the main C class, was not financially viable as it would require the installation of a new 24-volt electrical system plus associated alterations to the bodywork, and the cost of all this could not be justified as C 1 itself was due to be withdrawn when the CR vehicles now on order finally entered service. The engine exchange took place during November 1938 and C 1 was relicensed for service at Windsor as a petrol engined vehicle on 1st December.

In October 1938 a new, darker and much more hard-wearing shade of green paint was specified for all vehicles in the Country fleet in place of the mid-green previously used on the main side panels, and also on the roof and rear dome of the Cs and other more modern types of single decker. By March 1940 all 74 Cubs in the Country fleet had received the new livery, some as they went through their second routine overhaul and others at the start of the third overhaul cycle. At this time vehicles were still repainted at approximately eighteen-month intervals although war conditions were soon to bring this to an end.

During the early months of 1939, while the nation was preparing itself for war, three garages ceased operating one-man buses, allowing their Cubs to be transferred elsewhere. The first was Dorking whose only Cub departed for Leatherhead on 20th April. Next came Central Bus garage

The new, darker shade of green livery was introduced in 1938 and C 45 was one of the first Cubs to receive it. Note that the polished windscreens have now been painted over in the lighter shade of green to match the side windows. This vehicle, which had been new to Leatherhead on 14th June 1935, finally moved away to a new career at Dorking, and later Amersham, when Leatherhead lost the last of its Cubs on 1st July 1942. *Charles F Klapper © Omnibus Society*

C 10 wears the wartime livery of green and white, with the darker colour now covering the windscreen area. Its mudguard tips and bumpers are painted white in compliance with blackout requirements and the lamps are masked. Operated since new from Dunton Green garage, C 10 now nears the end of its operational life with London Transport. Made surplus by the conversion of route 421 to two-man operation using T types on 30th December 1942, it was withdrawn from service two days later. *Charles F Klapper © Omnibus Society*

Hanwell whose fleet of four red-liveried Cubs was replaced by larger crew-operated single deckers on 31st May and then sent to replace Darts at Enfield and Hornchurch. Watford High Street's conversion to crew operation was the largest affair, with its entire fleet of 9 Cubs and 1 Bedford moving out between 1st and 27th July. The most notable result of this exodus was that Epping garage received five Cubs in lieu of Bedfords, its first allocation of such vehicles since a lone example left there in July 1936.

The outbreak of war found the C class operating from 17 garages, the allocation of vehicles to each of these on 3rd September 1939 being as follows:

CENTRAL BUSES		
Enfield (E)	7	C 81, 86, 88-91, 94
Harrow Weald (HD)	6	C 77, 78, 87, 92, 93, 95
Mortlake (M)	2	C 80, 82
Hornchurch (RD)	8	C 51, 79, 83-85, 96-98
COUNTRY BUSES		
Addlestone (WY)	1	C 49
Amersham (MA)	10	C 29, 30, 33, 35, 36, 54, 61, 62, 71, 75
Chelsham (CM)	4	C 4, 7, 32, 66
Crawley (CY)	2	C 24, 28
Dunton Green (DG)	7	C 9, 10, 31, 53, 58, 59, 69
Epping (EP)	5	C 23, 63, 64, 73, 74
Guildford (GF)	6	C 34, 40-44
Hertford (HG)	11	C 12, 14-22, 60
Leatherhead (LH)	5	C 45-47, 52, 70
Northfleet (NF)	8	C 2, 6, 8, 11, 13, 65, 67, 68
St Albans (SA)	4	C 37-39, 72
Staines (ST)	3	C 25, 27, 48
Windsor (WR)	8	C 1, 3, 5, 26, 50, 55-57

With the outbreak of war the usual lighting restrictions to comply with blackout regulations were applied to the Cubs as were the white-painted mudguards, bumper and other markings, and before long the roofs of all vehicles had been painted grey as a camouflage measure. No time was lost in withdrawing operations which had absolutely no strategic value or were adequately covered by other services, and a 'Cub' service in the former category was Mortlake's sparsely patronised route 207 linking Barnes (Railway Tavern) and Richmond Park Golf Club. Two Cubs had been allocated to cover this one bus working, and on 16th September these were both transferred to Kingston in place of DAs, the first of the C class to operate there. They joined the first of the new rear-engined CRs which had been in service at Kingston from 9th September, and more of these were in course of delivery when the two Cubs arrived.

The only occasion when Cs were directly replaced by new CRs occurred at Windsor on 1st November 1939 when a small batch of six green-liveried ones arrived there. The displaced Cubs were mostly despatched to Hertford, enabling the last remaining Bedfords to be withdrawn from the fleet. Coincidentally, on the same day, on Central Buses, the last Dennis Dart was also withdrawn upon receipt of new CRs at Enfield.

One Windsor-based Cub was not transferred anywhere. Petrol-engined C 1 was delicensed, never again to run in service for London Transport. At a management meeting on 16th July 1940 it was agreed that C 1 could be sold to Banstead Urban District Council for £170 plus the value of its tyres, but a government embargo on bus sales prevented this from happening and C 1 remained in store, mostly at the yard in Forest Road, Walthamstow, for another 5½ years before being sold to London dealer Henry Lane of SW10 in January 1946. Subsequently, like a ghost from the past, it appeared on hire to London Transport during the early post-war bus shortage, firstly in the JM Motors fleet of J R Manuel of Holloway and later with Kentish Town-based Overland Lismore Coaches Ltd who finally pensioned if off in about June 1951.

The war years witnessed a gradual reduction in the requirement for 20-seaters in both the Central and Country Bus departments as increases in passenger loads demanded the use of larger, crew-operated vehicles. 1940 had barely begun when Epping lost the last of its Cubs on 3rd January as a belated result of the conversion of routes 339 and 339A to double deck operation using STLs on 13th December of the previous year. The Cubs had been little more than a short-term phenomenon at Epping, having been required there for barely six months. All five were reallocated for use elsewhere, but when Enfield's eight Cubs were all replaced by T class single deckers on routes 205 and 205A on 1st September 1940 they were not so lucky. No further use could immediately be found for them and all were put into store, some in the yard at Walthamstow and others at the Board's property in London Road, Norbiton. The inauguration on 20th November 1940 of new Cub route 238 (Emerson Park–Noak Hill) worked by Hornchurch garage did not call for any extra vehicles because of consequential alterations made to route 252. Although it could not have been known at the time, the 238 was destined to be the penultimate new Central Bus service ever to be inaugurated using Leyland Cubs.

An interesting feature of the war years was that various Cubs were allocated from time to time for miscellaneous duties, the details of many of which still remain unknown. Possibly some were used on staff shuttle duties, while others may have been covering short term essential contracts. Particularly unusual was the allocation of a Cub to New Cross tram depot, an arrangement which commenced on 12th November 1940 and which then lasted right through the war. Manned by a crew of two, its nightly nocturnal duty was to tour tram tracks looking for obstructions, in which role it was known locally as the 'Blitz Bus'. The first to be based at New Cross was C 88, followed by C 82, C 88 (again) and C 82 (again). Other unusual wartime allocations were:

Romford Green Line garage (now operationally closed) – C 26 (2/11/39-13/12/39)
Addlestone – C 16 (27/3/40-1/11/40)
Edgware (recorded as 'for use of paymaster') – C 88 (18/2/41-1/6/42)
Hemel Hempstead – C 30 (18/12/41-1/7/42)
Sidcup – C 94, later C 51 (15/11/43-4/3/44)
Old Kent Road – C 94 (30/8/44-1/10/44 and 19/12/44-1/2/45)

From about the start of 1940 a change in livery was applied to all Country area buses as they went through overhaul whereby the window surrounds formerly picked out in pale green were now in the same off-white as the Central Bus fleet. This was carried out when vehicles were overhauled, the programme for which, despite wartime restrictions, endeavoured to ensure that each vehicle went into one of the various overhaul shops at approximately two-yearly intervals. All the Country area Cubs had adopted the new livery by July 1941, and from

December 1941 the colour scheme was changed again when the grey roofs were replaced by brown.

Whereas 1941 was a fairly quiet year for the C class things turned out to be altogether different in 1942, especially in the second half of the year. The only occurrences of note in 1941 had been the conversion of a number of relatively minor one-man services to larger vehicles but in no cases did this involve any garages losing their entire Cub allocation. The summer programme on 2nd April, for instance, had seen the loss of Cub routes 329 and 389 at Hertford and 432 at Leatherhead, while the winter programme on 8th October included routes 373 and 398 at Amersham, 386 at Hertford, 416 at Leatherhead and 466 at Staines. In all cases the displaced Cubs were put into store. Their availability meant that, in April 1942, eleven vehicles were readily available for repainting into red livery to start a renewed life in the Central area, the first time vehicles of type 2C2 had ever operated there. From 11th April onwards C 14, 16, 19, 26, 31, 32, 44, 46, 49, 56 and 57 were repainted red, and on 1st May they were all relicensed and put into service at Uxbridge. London Transport's objective was to remove the entire CR class from service for the duration of the war. Unreliability and unpopularity with staff had spurred the decision to stop using them, which was hardly surprising at a time when the war was at its height and there were so many other distractions to deal with.

On the same day, 1st May 1942, most of Kingston's CRs were also removed from service but in this instance the incoming Cubs were all conventional Central Bus 1/2C2s. Six vehicles were received which had been in storage since Cubs were removed from Enfield on 1st September 1940. This still left a full allocation of CRs at Hounslow on route 237, but not for much longer. They were removed in two batches on 1st June and 1st July. On the first of these dates route 221 at Harrow Weald was converted to operation by T class vehicles, releasing five Cubs for use at Hounslow. The final three came from Kingston on 1st July, the day on which the Kingston–Staines portion of route 216 received single deck LTs to join the Ts already operating Kingston–Sunbury short workings that had been converted from Cub in January 1941.

In the Country area Epping garage commenced its third spell of Cub operation with the receipt of a single vehicle for route 399 on 6th May 1942. It did not stay long, however, as the route was taken over by T types on 1st July 1942. This was the date on which Leatherhead lost its final Cubs to nearby Dorking (which had fairly recently resumed one-man operation with a single ex-Leatherhead Cub on 15th April), and on the same date Staines garage also ceased one-man operation with the receipt of a T to replace its last remaining Cub. On 30th September 1942, which coincidentally was the day on which all Green Line services were withdrawn for the second time, Crawley lost its C allocation when circular service 426 was converted to double deck STL operation. On the same day Tunbridge Wells garage closed for operational purposes and an extra Cub was allocated to Dunton Green to shunt Tunbridge Wells staff from there to Sevenoaks bus station to work on revised Dunton Green garage rosters.

The autumn of 1942 witnessed two major retractions from one-man operation on Central Buses which resulted in its requirement for Cubs being reduced by more than two-thirds. The first came on 21st October at Uxbridge where the 10-bus allocation on route 223 was replaced by a smaller number of Scooter LTs, leaving only a much-reduced scheduled Cub requirement of three for route 224 for which four vehicles were retained. Uxbridge's entire fleet of Cubs consisted of ex-Country area 2C2s repainted red a few months earlier, and the decision was now taken to retain these in service and to delicense vehicles from the traditional 1/2C2 red fleet. The fears which the management held back in 1935 that the 2C2s would be unsuitable for Central Bus use had proved unfounded. In fact, their reliability had been thoroughly tested on the busy 223 from which they had emerged with flying colours. In contrast, the 4.7 litre indirect injection engines on the later vehicles had never proved to be really satisfactory despite considerable efforts to enhance their reliability. As a consequence, Cs displaced at Uxbridge were sent to Kingston, Hounslow and Hornchurch to replace vehicles from the original fleets at these three garages.

The next big conversion of the year occurred at Hounslow when the 237 was taken over by single deck LTs on 18th November 1942. Eight Hounslow Cubs were made redundant. Two 2C2s, which had only recently arrived from Uxbridge, were sent to Hornchurch and the remainder, which were all from the 1/2C2 batch, were delicensed. Thereafter a single Cub was retained at Hounslow purely for use as a staff bus in connection with Chiswick Works, where it was usually kept. No spare bus was available at Hounslow to cover for servicing or breakdowns, and whenever a replacement was needed it had to be borrowed from Uxbridge. The inroads into the 23-strong fleet of 1/2C2s meant that only seven of them remained in use at the end of 1942.

Wartime contraction in the use of 20-seaters was such that, by the close of 1942, 11 2C2s and 16 1/2C2s had been taken out of service along with the whole fleet of CRs. In view of the continuing trend towards the use of larger vehicles there was little prospect of the surplus Cubs being required for further use as service buses in the near future, if at all, so it was no surprise that London Transport responded favourably to an appeal by the Ministry of War Transport to lend some of them to other operators. In truth, the demand for 20-seaters elsewhere in the country was not great, but a start was made on 2nd December 1942 when C 16 was sent on loan to R S Braithwaite of 10 Yarm Lane, Stockton-on-Tees. This one returned to Chiswick nine months later, on 2nd September 1943, by which time a more substantive requirement had emerged from a small Gloucestershire operator in the Forest of Dean. W T Edwards & Sons of Joy's Green, Lydbrook was, in fact, expanding fast to meet wartime demands. A few local services, mostly of the market day variety, were worked from Lydbrook, but the main burden of the Company's work involved carrying factory staff and miners to their places of work, and it was mainly for this purpose that no fewer than seven Cubs were hired. C 91 and C 92 were relicensed by London Transport and despatched to Joy's Green in January 1943 followed by C 81 in February, C 90 and C 93 in March and, finally, C 52 and C 69 in December. C 93 was the first to be returned to London Transport on 18th March 1944 and the last was C 90 on 3rd May 1945. On their return to London all seven were immediately overhauled, the green ones at Reigate and the red ones at Chiswick tram depot, following which they were all put back into store.

An unusual occurrence was recorded on 15th February 1943 when three out of Windsor's four Cubs held for route 445 were replaced by relicensed red ones (C 79, 87, 89), followed by a fourth on 1st March (C 80). This was the first time that Cubs from the second batch had ever been allocated to Country Buses, and like much that happened during the war, the reason for it was not committed to paper and is now forgotten. There were plenty of green Cubs lying unused at the time, so the choice of these four vehicles for operation at Windsor was presumably for a specific reason and possibly something to do with their different mechanical specification. They were never repainted green while at Windsor, and their stay there ended on 30th August 1944 when route 445 was converted to STL operation, leaving the four red Cubs to be delicensed on 1st September, never to run for London Transport again.

Shortly before Windsor lost the last of its Cubs another Country Bus garage, St Albans, operated the class for the last time. Right from the start of London Transport days a number of 20-seaters had been based there, primarily for school work on behalf of Hertfordshire Education Committee but also running odd 'supplementary schedule' journeys on routes such as 304 and 391. On 1st July 1944 this arrangement finally came to an end when the last four Cubs at St Albans were replaced by larger vehicles.

The new arrangements at St Albans and Windsor marked the end of the wartime conversions from Cub operation to bigger vehicles. However, one more episode remains to be recorded. January 26th 1944 had witnessed the establishment of a new one-man route 225 (Eastcote–Northwood Hills) which, as it happened, was destined to be the last new route ever to be started on Central Buses using Cubs. Uxbridge garage received one additional Cub for it, the vehicle in question being red-liveried C 16 which was newly overhauled after returning off loan from Braithwaite at Stockton-on-Tees. This proved to be a very short-lived Cub operation as the 20-seater quickly proved inadequate and a T replaced it on 24th May 1944.

The fortunes of war had swung in favour of the allies when four days after D Day, on 10th June 1944, it was arranged that twelve surplus and currently unlicensed Cubs would be released to the Home Guard. Ten of them had been in store at Hemel Hempstead after being withdrawn from service, mostly in 1942, and the other two had been lying unused at Amersham. The full batch comprised C 12, 29, 33, 35, 51, 62, 71, 75, 83, 84, 85, 96 and they were despatched to a variety of locations. The arrangement lasted only just over six months and on 21st December 1944 all twelve returned for renewed storage, where they remained until sold in the following year.

The earlier summary of the allocation of Cubs to garages at the outbreak of the war on 3rd September 1939 showed 97 Cubs running from 17 garages. Such had been the level of attrition during the war that, on Victory in Europe Day on 8th May 1945, these totals had been reduced to 58 and 12 respectively as follows:

CENTRAL BUSES

Hornchurch (RD)	6	C 14, 19, 49, 57, 97, 98
Hounslow (AV)	1	C 44
Kingston (K)	4	C 31, 56, 78, 88
Uxbridge (UX)	4	C 16, 26, 32, 95

COUNTRY BUSES

Amersham (MA)	7	C 37-39, 41, 42, 50, 60
Chelsham (CM)	7	C 4, 7, 24, 47, 59, 61, 66
Dorking (DS)	2	C 45, 70
Dunton Green (DG)	6	C 36, 53, 58, 63, 64, 73
Guildford (GF)	6	C 22, 23, 25, 34, 40, 69
Hertford (HG)	8	C 3, 15, 17, 18, 20, 21, 30, 74
Northfleet (NF)	6	C 2, 8, 11, 13, 28, 65

TRAM & TROLLEYBUS DEPT

New Cross	1	C 82

Peace had not long returned to Europe when, in August 1945, the Ministry of War Transport wrote to London Transport on behalf of the European Inland Transport Organisation to enquire as to the availability of single deck buses which could be released by the Board for use in continental Europe, either through direct sales or on a leasing arrangement. London Transport reviewed the request quickly and sympathetically, but came to the conclusion that the only buses in the fleet which might be regarded as surplus to immediate or possible future requirements were 20-seaters, namely 39 C-type and 48 CRs. The latter were deemed unsuitable for export because of the difficulty in obtaining spares and what was tactfully described as "certain technical reasons", but on 21st August it was decided that 30 of the C-type could be released on the understanding that they would be sold rather than leased. In due course arrangements were concluded with the Belgian Economic Mission of 115 Eaton Square SW1 to take all 30 at a price of £650 each.

Many of the Cubs exported to Belgium in 1945 were worked hard by their new owners in the difficult early post-war years. Minus some of its glass and other bits and pieces, the former C 29 was working for Arseen Govaert-Miny when photographed in Ostend on the high frequency local service to Conterdam. The photograph is believed to have been taken in the summer of 1951 and the vehicle carries the symbol of Belgian National Railways on whose behalf the service was operated. *John L Smith*

Although out of use, C 54 was still in existence in Belgium and apparently virtually intact as late as April 1962. Photographed opposite the garage of its owner Paul Pironet of Pepinster, near Verviers, it had run between there and Trooz under the title of "Les Autobus de la Vallée de la Vesdre". A fold-down step appears to have been inserted into the original emergency door which would have served as the main entrance and exit point during its time in Belgium, but otherwise the bodywork looks virtually unchanged from its London Transport days. *John C Gillham © Tony Peters*

On 26th and 27th November 1945 the 30 Cs left London Transport ownership for new lives in various parts of Belgium, usually (or perhaps entirely) in the ownership of small private operators but often working services under the oversight of the state-owned Belgian National Railways. The 30 vehicles sold were:

Green – C 5, 6, 9, 10, 12, 27, 29, 33, 35, 43, 48, 54, 55, 62, 67, 68, 71, 72, 75
Red – C 46, 51, 79, 80, 83-87, 89, 96

Meanwhile, behind the scenes a programme was quietly under way to remove the unsatisfactory indirect injection engines from Cubs that were likely to remain in service for any reasonable length of time and to replace them with the type of engine that powered the C 2-75 batch. It is not known when individual vehicles were dealt with, although it is thought – but not positively established – that the programme may have commenced as far back as November 1942 when Cs 46 and 94 at Uxbridge exchanged engines and C 46 had ceased to be operational. Certainly, by January 1951 eleven had been dealt with (C 77, 78, 81, 82, 90-95, 98). In that month authority was given to permanently withdraw the only two vehicles still complying to 1/2C standards; these were C 88 and C 97, which last ran at Chelsham in January 1951 and Guildford in August 1950 respectively. In March 1951 their bodies were burnt at Chiswick and their chassis dismantled to salvage usable spare parts.

Barely had the war come to an end when one-man operation ceased at Uxbridge, leaving Hornchurch and Kingston as the only Central Bus garages running Cubs on scheduled services. T-type vehicles had taken over route 224 on 17th October 1945, a week later than originally planned, and at the same time the single C providing staff bus cover based at Hounslow was also withdrawn. This still left a Cub at New Cross which finally left there on 22nd July 1946. Kingston was the next garage scheduled to follow Uxbridge's example with the conversion of route 206 to Ts on 9th October 1946, but in fact the replacement buses were available earlier than originally anticipated and Kingston's last Cubs moved north to Forest Gate garage on 11th September to start a new phase in their operational lives.

Who would have thought, when C 83 was new in April 1936, that less than a decade later it would end up operating in a war-ravaged Belgium? C 83 was the first of three new 1/2C2s allocated to Barking where they joined the experimental-engined C 51 on Romford local services. It was obviously a hot summer day in Romford when C 83 was seen on the 252 with most of its windows wide open.
J F Higham © A B Cross

Photographed late in the war, and now with a brown roof instead of grey, C 47 was one of seven Cs kept at Chelsham, mainly for use on the 464/465/485 group of routes linking Westerham with Edenbridge. The sometimes hit-and-miss application of anti-shatter mesh to protect the window glass is evidenced here where it is visible only on one window. C 47 was still at Chelsham at the war's end and stayed there until it was required elsewhere in January 1950.

The war had taken its toll on London Transport's bus fleet which had sunk into a precarious condition. With a high proportion of vehicles time expired, a huge accumulation of maintenance arrears, and a totally inadequate supply of new buses, the Board's inability to provide an adequate and reliable service was a problem that was to persist virtually to the end of the decade. The only section of the fleet where a surplus of rolling stock existed was in 20-seaters, and despite their total unsuitability the decision was taken to retrieve all the unused ones out of storage and to relicense them for peak hour work as crew-operated vehicles to help relieve the pressure on key Central Bus services. On 15th June 1946 many such vehicles were relicensed and distributed to garages ready to start work on Monday the 17th, much to the amazement of many long-suffering members of the travelling public. Most of the vehicles relicensed were CRs as these represented the majority of 20-seaters in store, but a few red-liveried Cs were also evident, initially at Barking garage with an allocation for route 23B. Forest Gate garage also received five in September followed by three at Tottenham in October and a single one at Streatham in November, in the latter two instances supplementing a much larger CR allocation. A considerable number of subsequent reallocations between these garages took place to suit operational needs and the last use of Cs on Central Bus crew work appears to have taken place on Tuesday 2nd March 1948 after which the last remaining one, C 88, was transferred from Barking to join the conventional one-man fleet at Hornchurch.

With peace firmly established the Country Bus & Coach department was able to take stock of its

C 1 disappeared from public view on 1st November 1939 only to reappear, largely unaltered but now with a later style radiator, in the colours of J R Manuel's JM Motors of Holloway almost six years later, in about June 1946. London Transport's arrangement with the Passenger Vehicle Operators' Association to hire coaches and their drivers from 29th October 1947, with the Board providing conductors and fuel, found C 1 unexpectedly back on London bus service work. In this instance it is seen working route 19 from Holloway garage. C 1's oddly sited blind winding handle is still very much in evidence albeit out of use, a printed sticker having been attached to the destination box glass.

situation and in the process it produced a list of four new one-man operated services of a developmental nature that it wished to introduce subject to the release of suitable vehicles by Central Buses. This list was drawn up towards the end of 1946 when the use of 20-seaters as emergency cover on Central area services was at its height, and comprised:

372	Hertford–Welwyn Garden City
383	Hitchin–Weston
479	Dartford–Farningham
494	Oxted–East Grinstead

The department's hope was that these four services could be commenced during 1947 to avoid being faced with applications for similar operations from independent operators, but the latter were facing similar problems of rolling stock shortage and no applications came from the private sector. The first of these new operations did not finally get under way until 7th January 1948 by which time the London Passenger Transport Board had ceased to exist. Under the Labour government's nationalisation programme the new London Transport Executive had taken over on 1st January 1948 and it was under its auspices that East Grinstead garage began running route 494 using red C 31 transferred in from Forest Gate. Route 383 came next on 18th February using two red Cubs from Hitchin garage, followed by the 372 on 19th May for which Hertford received an additional Cub. A pause of more than a year then ensued before the programme was completed with Swanley garage receiving its first-ever Cubs for the start of route 479 on 15th June 1949. A further new service supplementary to the original list was 393 (Harlow–Hoddesdon) for which Epping garage returned yet again to the list of Cub users, also on 15th June.

In a negative vein, Tuesday 19th July 1949 saw the last day of one-man operation on Central Buses after which routes 238 and 252 at Hornchurch were turned over to larger T class vehicles. A generous allocation of nine Cs had latterly been provided to cover the six scheduled workings there. Four of these went immediately to the Country area while the other five were delicensed, although all were used again later. The last nine Central Bus Cubs were C 77, 78, 81, 88, 90-92, 95, 98.

Red Cubs were no longer needed and all of the Hornchurch vehicles except C 88 were later repainted green. In fact, repaints from red to green had commenced as early as December 1946 and this process continued spasmodically through to 1952 (and possibly even 1953 in the case of C 90). The vehicles dealt with were:

C 56	December 1946	C 31	January 1950
C 49	January 1947	C 78	April 1950
C 16	April 1947	C 93	May 1950
C 44	December 1947	C 77	June 1950
C 57	April 1948	C 94	June 1950
C 32	July 1948	C 95	September 1950
C 26	August 1948	C 14	December 1951
C 97	April 1949	C 92	February 1952
C 19	May 1949	C 81	September 1952
C 98	July 1949	C 90	date unknown
C 82	August 1949		

In addition, C 91, which was serving at the time as a staff bus, was repainted from red into an all-over grey livery in March 1950 but it subsequently became green in January 1952.

Still on the subject of livery, all Country Bus Cubs, when repainted, received standard green and white until the spring of 1948 when the department adopted a revised styling for certain classes of vehicles, including the C class, and also the 4Q4s, 9T9s, 11T11s and post-war double deck STLs. The relief colour was changed to cream instead of white, but this was applied only around the passenger windows leaving the cab side windows, and also the entire fronts where applicable, completely green. This livery change proved short lived, lasting only from May to about the end of 1948, after which the old styling was adopted once again and it was received by only a few Cubs. On 31st August 1950 the very last Cub to receive a full overhaul, C 95, emerged from Chiswick in the latest 'all green' livery on which the entire fleet (except Green Line coaches) was now being standardised for spray painting purposes, with cream reliefs on single deckers restricted to narrow mouldings above and below the windows. Subsequent to this, at least four others (C 18, 23, 58, 90) received this same treatment when their bodies were deemed to need repainting in about 1952 or early 1953.

Cubs were allocated to East Grinstead garage for the first time in January 1948 when route 494 opened up new bus territory over previously unserved roads between Lingfield and Oxted. Standing in the small bus station at East Grinstead, the highest numbered 20-seater C 98 was a later newcomer to the service after being repainted from red to green in July 1949. It left when route 494 was turned over to CRs in September 1951. *D A Jones*

When the new decade arrived the Country area Cubs were still more than adequately fulfilling their role on a wide miscellany of services from eleven garages. However, these sturdy vehicles had now long exceeded their intended life span and were looking very out-dated, and it was generally accepted that their end could not be far away. Indeed, the decision had already been taken at Executive level back in September 1949 that approval should be obtained urgently for their replacement if this was to be achieved, as was then hoped, in 1951.

At one stage an alternative to complete replacement was mooted whereby the Cubs' running units – which were still functioning well – could be removed and re-used within new body structures, a procedure which J B Beadle at Dartford had pioneered with some success for several Tilling group companies also within the nationalised sector. London Transport had shown sufficient interest in the project to examine one of Beadle's chassisless vehicles at Chiswick on 10th December 1948, but consideration of this novel approach was abruptly halted when Leyland announced its intention to cease supplying spare parts for Cubs.

Meanwhile, there still remained a few new rural services in areas not currently served that London Transport still wished to inaugurate with 20-seaters. After having sold so many Cs in 1945 the situation with regard to availability was becoming 'tight', to the extent that it had reluctantly become necessary to augment them with a few CRs, a process which had begun at East Grinstead in August 1948, spreading to Chelsham in November of that year. Both garages received two CRs apiece, but neither lost is entire allocation of Cs and, after a few months of spasmodic operation, the use of CRs ceased in July 1949. Of necessity, it had to resume in September 1951, this time at East Grinstead (for the second time) and Epping.

After Mr Manuel sold C 1 to Overland Lismore Coaches of Kentish Town it stayed on London bus service and could normally be found working route 68 from Chalk Farm garage. When this work ceased an effort was made to make it look more coach-like, enabling it to continue earning a living up to about June 1951. In this view C 1 looks well cared for with a newly re-plated front bumper and all its other bright metal work shining. *Alan B Cross*

Occasional instances of Cubs appearing on conductor-worked Central Bus services could be found until as late as 1949. In this instance Hornchurch's C 82 is working the busy 175 along with one of the normal residents of the route, Massey-bodied G 186, on 28th April 1949. Almost four years after the war ended it still carries masking on its side lights and, even more of a rarity at such a comparatively late stage in its working life, it still displays chromium plated windscreen surrounds. *Alan B Cross*

The takeover of routes 238 and 252 by larger vehicles on 20th July 1949 heralded a gap of 15 years in one man operation on Central Buses, which was not resumed until November 1964 using RFs. It was still business as usual when C 81 was photographed on route 238 on 9th July with just over a week still to go. After a spell as a staff bus, C 81 was repainted green and re-entered service at Epping on 1st September 1952. *Alan B Cross*

C 31 stands in Hertford bus station on route 372, a new Cub operation introduced in May 1948. Still carrying red livery after being made redundant at Kingston in September 1946, it has since operated at Forest Gate and East Grinstead before arriving at Hertford in October 1948. One of a number of St Albans vehicles relieved of their front bumpers in the mid-1930s, it is still in this state and remained so until withdrawal even though it was later repainted back to green livery in January 1950.
D A Thompson

C 26 was one of a small number of Cubs repainted in the short-lived 1948 green and cream livery style on which only the windows of the passenger areas were picked out in the lighter colour. In this instance the repainting was carried out during its final overhaul which occupied eleven weeks between June and August 1948 and included a colour change from red to green. Seen at Orpington station on the busy 471 Pratts Bottom circular, it is operating from Dunton Green garage, its base from April 1946 until withdrawal in November 1953. *J H Aston*

Most unusually for such elderly vehicles two Cubs, including C 94, were equipped with decorative rear wheel discs at overhaul in June 1950 (the other was C 25). Additionally it was very rare for one of the second batch to be provided with a front bumper, but these were both features that C 94 still retained when photographed at Hitchin in early 1952. The wheel discs had gone by the time it reached Northfleet garage in May 1953 but the bumper survived until the end. This is one of the Cubs which, happily, survives in preservation to the present day.
Alan B Cross

The next new Cub service commenced on 1st March 1950 when Dorking received an extra C for the 449 (Dorking–Ewhurst). This was followed by the 336A (Rickmansworth–Loudwater). Strictly speaking this was not a new service, having been operated since 1928 by the housing developer Land & Estates Ltd. This company now wished to cease its bus activities and, with no-one else willing to take responsibility, London Transport agreed to provide a replacement, albeit very reluctantly as it was expected to produce a loss of £1,370 per annum. The former 'Troutstream Bus', as it was known, operated for the last time on Tuesday 6th June 1950 and the 336A took over on the next day with newly-overhauled C 77 providing the rolling stock and the former Troutstream driver Harry Simmonds at the wheel. Although officially allocated to Leavesden Road garage – the first Cub to run from there – the bus was outstationed in the old coach house at Loudwater and only returned to Leavesden Road on Saturday afternoons for servicing.

A misty day at Dorking garage-cum-bus station in March 1951 finds both of this garage's Cub routes coming together with C 70 on the old-established 433 and C 47 on newcomer route 449 introduced in March 1950. Also in this picture is RT 634 from Leatherhead garage on the trunk service 470 with, just visible in the background, a Green Line TF. Both Cubs remained at Dorking until the end came on 1st December 1953 and subsequently worked together again for the South Western Bus Company in Ceylon. *Alan B Cross*

C 95 was one of the final Cub contingent at Hornchurch in July 1949, and after a period in store it became the last of the class to be overhauled and the first to appear in the 'all green' livery when relicensed at Guildford on 1st September 1950. Photographed at the Ewhurst terminus of route 448, it presents a fairly gloomy sight especially as the radiator shell is also painted green. *Ken Blacker*

The final new Cub venture of 1950 was route 381 (Epping-Toothill) worked from Epping garage which began on 30th August. In 1951 came the introduction of an experimental service 315 between Welwyn Garden City and Wheathampstead. In line with its prime intention of acting as a service for tourists, it began operation on Good Friday 23rd March employing a single Cub from Hatfield garage, its first and last allocation of a vehicle of this type. The 315 was primarily intended to serve the village of Ayot St Lawrence and, in particular, the former home of the renowned author George Bernard Shaw who had died in November 1950. Sadly, the 315 proved to be a commercial disaster and ran for the last time on 12th August 1951. This left just one more new operation to be inaugurated by a Cub, route 387 linking Tring and Aldbury. The 387 itself was by no means new, being a well-established operation on which a one-man C replaced a crewed 4Q4 as an economy measure. This arrangement officially commenced with the start of the winter programme on 26th September 1951 but Tring garage did not receive an allocation of Cubs until the first week of October in the form of one red one (C 90) and one green (C 65), sufficient to cover the scheduled service and to provide a spare.

1953 marked the end of the road for most of the one-man Cubs. Two of them (C 52, 64) were withdrawn from service on 1st June and almost immediately broken up at Chiswick to provide spare parts to keep the others going. Construction of the new GSs, though much belated, was now under way, and a programme was drawn up to determine the order in which they would be allocated to garages with Hitchin, for some unknown reason, pinpointed as top priority. In the end, however, Chelsham was the first to put GSs into service on 26th October, a few days before Hitchin on 1st November. This was the date on which the last CRs were taken out of service, so it marked a temporary return of Cs to East Grinstead and Tring, both of which had been without them for a while, having relied entirely on CRs from 1st October 1951 and 1st December 1952 respectively.

Few experimental services tried out by London Transport proved as instantly and deeply unremunerative as route 315, but at least it gave Hatfield garage its one and only opportunity to operate a Cub, albeit only from 23rd March to 31st August 1951. The vehicle selected to operate it was C 23, one of the most recent in the fleet to be overhauled (in August 1950). It was photographed at the Welwyn Garden City terminus in the company of RT 1088, also a Hatfield bus. *R K Blencowe*

The Cub route that never was! Route 481 was a new operation in Epsom introduced on 6th May 1953 and scheduled from the outset for operation by CRs . Although Leatherhead garage was sufficiently well stocked to cover the operation, the dubious reliability of the CRs meant that a C sometimes had to be borrowed which, in this instance is C 53 from Dunton Green. CR 43, standing behind, illustrates the height difference between the two types of 20-seaters which results in the CR appearing to be by far the more bulky of the two. *Alan Nightingale collection*

C 7 was one of Swanley's last two Cubs, both of which were received for the introduction of route 479 in June 1949 and were still in service there until final withdrawal on 16th November 1953. Note that it has a divided rear window which was by no means a rare feature on the Cs in post war years, there having been at least seven recorded instances besides C 7. *F G Reynolds*

Once they started arriving, the new GSs were placed into service thick and fast as follows:

Garage	*Date of 1st GS allocation*	*Day when final C delicensed*
Chelsham (CM)	26/10/53	1/11/53
Hitchin (HN)	1/11/53	1/11/53
Hertford (HG)	about 4/11/53	9/11/53
Swanley (SJ)	14/11/53	16/11/53
Dunton Green (DG)	16/11/53	1/12/53
Guildford (GF)	27/11/53	1/12/53
Dorking (DS)	1/12/53	1/12/53
Epping (EP)	1/12/53	1/12/53
Northfleet (NF)	9/12/53	1/1/54
East Grinstead (EG)	16/12/53	1/1/54
Amersham (MA)	21/12/53	1/1/54
Tring (TG)	1/1/54	1/1/54

So, 1954 began with just one C left in service. This was C 60 which, since replacing C 77 in November 1950, had been the regular vehicle working estate service 336A to Loudwater. Since 18th June 1952 C 60 had been the responsibility of the new Garston garage as a result of the closure of Leavesden Road, although it was still outstationed for six nights a week in the old coach house at Loudwater. Although this building could comfortably accommodate a Cub, the much taller GS would not fit inside until the doorway had been suitably heightened. News finally came through on 13th April 1954 that this work had now been done but it was decided that, in order to save the licence duty, C 60 should remain in service until the end of the month with a GS taking up running from 1st May.

On Friday 30th April 1954 C 60 duly made its last service runs, exactly four months after the demise of the rest of the class and a massive 13 years later than the six years life originally projected for it. With its last run completed, C 60 was immediately driven to Garston garage and delicensed with a GS being delivered to Loudwater ready for the next day's operation.

Old and new parked up outside Dartford garage late in 1953 at some time between 14th November when Swanley started running GSs and their introduction at Northfleet on 9th December. The new GS class makes a fitting and handsome successor to the Cubs, but the latter are characterful little workhorses and will be missed. C 66 spent almost the entirety of its post-war working life at Northfleet. *Alan Nightingale collection*

C 60, the Loudwater outstationed bus, outlived all the other Cubs in passenger service by a full four months, and even at the end it still had a very presentable appearance. Having worked on the 336A from November 1950, it passed from being the responsibility of Leavesden Road garage in June 1952 to Garston whose garage identification plates it displays while waiting on the bus stand in Rickmansworth. *F G Reynolds*

After the demise of C 60 the joy of riding on the little Leyland Cubs was not entirely lost, although henceforth it could only be enjoyed by London Transport staff. Since the war there had grown a tradition of using the occasional Cub as a staff bus, a notable one being C 81 which had been officially allocated to Hounslow (although it was usually housed overnight in Isleworth trolleybus depot) for use by Chiswick staff from November 1950 to July 1952. The last staff bus Cubs of all were Cs 25 and 82 which were based at Abbey Wood and Plumstead respectively to transport displaced Charlton Works personnel to other work locations, which they did until they were finally delicensed on 25th March 1955. Even this did not quite mark the end for these characterful and highly successful little vehicles for C 25 was loaned to the Motor Industry Research Association at Nuneaton from 26th April for a fortnight. Its return to Chiswick on 11th May finally brought the story to an end.

Sales of the now-redundant Cubs commenced in August 1954, and despite their age and obsolescence the majority were snapped up for further use. Some inevitably went to building contractors for carrying workers to construction sites and a few went to welfare organisations, such as C 38 to the St John Ambulance Brigade in Cricklewood and C 68 to the Central Middlesex Hospital. However, the bulk of sales, embracing 30 vehicles in all, were exported via W North, the Leeds dealer, to the South Western Omnibus Company in Ceylon (nowadays known as Sri Lanka). South Western was the largest operator on the island and it had a large garage and extensive workshops at Ratmalana, a few miles south of the capital, Colombo. Purchased by the Company in January 1955, all 30 underwent a miraculous transformation whereby they were lengthened to allow an extra bay to be added. All were from the Short Brothers batch, permitting commonality of body parts, many of which may well have been obtained by North's from vehicles that they had broken up. Now capable of accommodating 33 seats plus another eight standing, they were visibly unaltered externally except for their extra length and a narrowing of the passenger doorway, and all 30 gave good service in Ceylon in their new guise. They passed to the ownership of the Ceylon Transport Board when the whole of the island's bus network was nationalised on 1st January 1958 and could still be found running, in increasingly dwindling numbers, until the autumn of 1965. For the record, the Ceylon 33-seaters were the former C 2, 3, 8, 11, 13-16, 18-20, 22-24, 28, 32, 34, 37, 45, 47, 49, 56, 58, 60, 61, 65, 66, 70, 73, 74.

Of the 62 Cs that remained in service right up to the start of the GS era in October 1953 all except ten saw further use in one form or another. The main user of them in the UK was A G Linfield Ltd of Thakeham, Sussex, a huge agricultural concern and Europe's largest mushroom grower, who purchased six Cubs via W North of Leeds for transporting its staff of more than 600 employees. C 53, which was photographed in Worthing in July 1958, remained in Linfield's service until about June 1962. *John C Gillham © Tony Peters*

The Cubs, in their locally extended form, gave good service in Ceylon, firstly with the South Western Omnibus Company and then with the nationalised Ceylon Transport Board. A unidentified C stands with three other Leylands, an ex-London Transport CR and two Leyland Comets purchased new, in South Western's spacious depot yard at Ratmalana. Note the modified doorway on the Cub. *British Commercial Vehicle Museum*

The joys of preservation! The excitement of finding a historic gem to preserve can sometimes pall when the reality of the work involved becomes apparent. C 94 had been used by Harry Russell Ltd of Woodley, Reading before falling derelict still with its Northfleet garage destination blind in place albeit a little tattered. Its new owner, Alan Cross (whose photographic skills have proved invaluable to many historians) shovels the snow away to study the vehicle's interior in January 1963. *Ken Blacker*

More than six decades after they last ran for London Transport, two of the 20-seat Cubs still survive thanks to dedicated efforts of private preservationists over the ensuing years, one vehicle from each batch. C 4, resplendently restored into its original Country Bus livery, is now in the Ensign collection and C 94, in Central Bus red, is now in the London Transport Museum's reserve collection at Acton.

After extensive restoration work, C 94 is now in the ownership of the London Transport Museum under whose sponsorship it is seen participating in a Historic Commercial Vehicle Society London to Brighton run in recent times. Appropriately, as a one-time Central Bus vehicle, C 94 carries a close approximation to the livery which it wore in early post-war years although the windscreen surrounds would then have been picked out in red. C 4 has been beautifully restored by Ensign into its original pre-war livery, over six decades after withdrawal from Chelsham in October 1953. *Michael H C Baker*

THE INTER STATION CUBS C 106 – C 113

Following hot on the heels of the last of the Central Bus Cubs came the first of a small batch of vehicles which were possibly the oddest and most distinctive buses ever purchased by London Transport. C 106-C 113 were the famous Inter Station Cubs.

The gap in the fleet numbering series between C 98 and C 106 occurred because the intervening numbers were used for a batch of Leyland Cub goods vehicles bearing the numbers C 99-105, and there was also one at the end numbered C 114. These formed part of a fleet of 29 Cub lorries and vans purchased by the Board during 1936 for a variety of purposes, all of which were petrol engined forward control vehicles of Leyland type SKZ1. London Transport practice at the time was to allow each major department to number its miscellaneous vehicles in a batch of its own, so it was only the eight Cubs controlled within the Chiswick regime which gained C numbers. They served a variety of purposes, C 99-101 being open lorries attached to the Building Department at Parsons Green, C 102-105 were stores lorries based at Chiswick and carrying various types of body, while C 114 was a closed catering van kept at Lillie Bridge. In appearance these vehicles bore no outward resemblance to London Transport's Cub passenger vehicles, having entirely different radiators. Each carried a small letter L – standing for Lorry – behind the fleet number (e.g. C 102L) in the same way that Country area buses were distinguished by the letter B and coaches by the letter C. C 114 was the last of its type to be purchased by London Transport, which turned to Albion for its 1937 lorry deliveries, and in 1939 the use of the C feet numbers ceased when London Transport renumbered all its miscellaneous vehicles into a single series and the Cubs were blocked together as 180C to 208C.

The Inter Station bus service for which C 106-113 were purchased could be traced back to 1929

With their highly distinctive shape and vivid livery the eight Inter Station Cubs could hardly fail to be noticed. In a manufacturer's photograph, newly completed C 108 is seen before receiving its registration plates. Its licence was issued on 1st October 1936, eight days ahead of the formal service introduction. A surprising feature of the body design was that no destination displays were provided at the front, and what looks superficially like a blind box above the windscreen is, in fact, a push-out ventilator for the driver's cab.

and had been operated by P Hearn under contract from the main line railways using Daimler coaches. In 1935 London Transport obtained a contract to take over from Hearn starting on 1st July 1936, and it was determined from the outset to make it an eye-catching, prestigious operation using vehicles that would attract the maximum attention wherever they ran. A fairly frequent all-day service was planned which was aimed principally at passengers travelling to and from the continent. As these traditionally had large quantities of accompanying luggage it was thought essential to provide specialised rolling stock specially equipped to cope with it. The Chiswick drawing office set about designing a half-deck type of vehicle with a rear 'observation' saloon mounted above a huge luggage compartment which would be accessed by a pair of tall outward opening doors at the back with a smaller one on each side. By the end of 1935 their design ideas had been circulated and approved by the General Managers of all the main line railway companies.

Early in 1936 the purchase of eight Leyland Cub chassis was authorised, and Park Royal was subsequently chosen to build the bodies, all within a cost ceiling of £11,200. This equated to a very high price per vehicle of £1,400 and was indicative of the inflated cost of building to a specialised design. In fact, the final cost worked out at £1,428 per vehicle, far in excess of what had been paid for conventionally-bodied Cubs such as C 77-98 which had cost £1,070 apiece. Such was the urgency for the construction of the new vehicles that Leyland was asked to begin the project immediately, ahead of the official order being placed which would be confirmed to them later.

The eight Cub chassis – which were the first buses ever to be purchased by London Transport outside its normal rolling stock replacement programme – differed in several aspects from the Cubs already purchased. They were forward control rather than bonneted and were equipped with petrol engines. The latter feature was thought to be desirable having regard to the necessity of manoeuvring within the confines of several main line stations where it was possible there might be objections to the odour of oil engine exhaust. The petrol engine, a Z-type unit with 3½ins bore and 5ins stroke rated at 4.7 litres, was quite unlike the side valve unit fitted to C 1. The Z-type was a modern overhead valve engine which had a reputation for being lively, smooth running and reliable. The braking system, clutch and gearbox were the same as on the Board's earlier Cubs and the same rear axle ratio of 6.5:1 was employed, but a point of difference was that the fuel tank was now situated on the nearside.

An offside view of the lowest numbered of the batch, C 106, shows the central position of the full-depth emergency exit door. A year into its service career, in August 1937, C 106 stands at Eccleston Bridge, Victoria awaiting the next leg of its journey. It is obviously a hot day and most windows are fully open including those at the front of the observation saloon which would have ensured a very breezy ride for the passengers sitting inside. Even the driver's cab ventilator is wide open. *D W K Jones*

A view inside the spacious luggage boot of a still to be registered vehicle just after completion at the Park Royal factory shows that even the wheel arches have been given squared tops to provide maximum storage space. *Park Royal Vehicles*

The unusual design made it difficult to obtain a good photographic overview of the internal layout, but this view manages to illustrate how the seating for 20 passengers was arranged using standard Green Line style seats. The steps up to the seats in the observation saloon, plus the accompanying handrail arrangements, can also be clearly seen. *London Transport Museum*

The chassis was Leyland's model SKPZ2 and the eight vehicles were delivered with chassis numbers 3130-3137.

Before the order was finally confirmed and the approved body plans were despatched to Park Royal a number of the original design features had already been modified. These included modifying the slope of the front windows of the observation saloon to bring them in line with the contour of the driver's cab windscreen, and a revision to the staircase and handrails to provide better circulating space just inside the doorway. Park Royal received the first chassis, C 106, on 28th April 1936 and work was put in hand almost straight away to construct a body to act as a guinea pig for the following seven. Its date of completion is not known, but C 106 was licensed on 26th June 1936, officially at Old Kent Road garage from which the Inter Station service was destined to operate, but it was more often to be found in the early days at Chiswick where it was thoroughly tested, and sometimes back at Park Royal for modifications to be made.

The completed vehicle was certainly eye-catching, not just because of its shape but also by virtue of its striking livery officially described as pale blue and primrose, these colours being very effectively offset by black horizontal mouldings, roof, mudguards and wheels. Although destined to be crew worked, the seating capacity was only 20 with eight seats in the lower saloon and twelve on the observation deck. Two quite steep steps led up from the main saloon floor to the upper level, where a further step up was needed to reach seat level. The main passenger entrance was in the centre of the nearside enclosed by a door which slid frontwards, and the hinged emergency exit

A trio of Inter Station Cubs at Waterloo station on 11th June 1937 with the crew of C 109 ushering passengers on board and stowing luggage in the boot of the coach. Although labelled externally for emergency use, the offside door also served as a convenient point of entry at stops where offside loading applied. *London Transport Museum*

The wartime resumption of Inter Station operation on 15th December 1943 provided a convenient link between main line stations at a flat fare of one shilling for passengers unable to find taxis. British troops are seen boarding at Waterloo five days after the service recommenced. *Topical Press*

door was immediately opposite this on the offside. Ventilation was provided by means of three half-drop windows in the lower section and four above, while the front windows of the upper saloon could be opened outwards and there were also two fixed roof ventilators in the upper section. Internal décor was provided by the Board's favoured rexine covering to all the window surrounds and side panels and the latest style of tubular-framed lightweight seats was used with cushions covered in Green Line style moquette. With an unladen weight of 4ton 18cwt the Inter Station Cubs were 1ton 3cwt heavier than C 77-98 despite their petrol engines being some 2cwt lighter. They shared the same 15ft 6ins wheelbase as the earlier vehicles and were almost the same length at 23ft 11½ins. The biggest difference was, of course, that they were considerably taller at 10ft 6ins compared to 8ft 10ins of the conventionally-bodied Cubs.

C 106–113 were registered CLX 543-550, carried body numbers 16722-16729 and given the type classification 3C3. The commencement by London Transport of the Inter Station service did not take place on the planned date of 1st July 1936 but was deferred, because of late delivery of the new vehicles, to 7th October. C 106, 108-112 were available at Old Kent Road for the start of the operation and were joined by C 107 and C 113 on 21st October. Prior to this, all eight had been sent back to Park Royal at least once and sometimes twice for rectification work or modifications to be carried out including to the wheel arches.

On 8th November 1944 C 111 was the first of the class to be allocated for use by Railway Air Services whose name was displayed on the otherwise unaltered Inter Station livery. To comply with blackout requirements the usual white wartime markings have been applied while replacement headlamps have been fitted and the fog lamp lowered to sit below the bumper. *Ken Blacker collection*

The Inter Station service ran successfully and uneventfully until the outbreak of war when it became an early casualty. On 16th December 1939 it ceased and the eight Cubs were delicensed. Initially they were all taken to Crawley garage for storage and then, in July 1940, to the Green Line garage at Romford. Subsequently they were dispersed, probably for safe keeping during the Blitz away from the possibility of bomb damage, to a variety of Country Bus locations before being reunited back at Romford by October 1942.

Quite early on in the war the future of the eight vehicles came under official scrutiny. With only 20 seats and unsuitable for one-man operation, they were deemed too small to be of use in ordinary service, and a suggestion that their bodies should be removed and put into storage and the chassis re-used as lorries was not pursued when it was found that there was no demand for this type of chassis on war work. Conversion for use as armoured vehicles, as had occurred with some ST chassis, was also ruled out. A June 1940 investigation into turning them into demonstration kitchens for the Ministry of Food also came to nothing. Yet another suggestion was that they might be used to supplement the Green Line coach network if more Green Line coaches had to be converted to ambulances, but the need for this did not arise. Selling them was also considered, but when this was informally discussed with the four main line railway companies in August 1940 they were all firmly opposed to the idea in case the Inter Station service was needed in future. It was finally decided that all that could be done was to keep them in safe storage for the duration of the war.

A short-term use for C 113 and C 111 was found in April and May 1941 respectively when they were taken to Chiswick Works and licensed as goods vehicles for London Aircraft Production for whom the works served as a major manufacturer of components. They were repainted into service vehicle green and continued to carry this livery when relicensed as Hackney vehicles in July 1941, presumably to convey LAP staff. They relinquished this role in November 1941 and were placed back into storage pending their next role. C 112 was used temporarily by the Home Guard at Luton between 13th July and 30th October 1942, and towards the end of that year a more substantial role was found for the whole batch. Between 15th December and 7th January 1943, all eight were despatched on loan to ENSA (Entertainments National Service Association) which famously provided entertainment to troops on the battlefield and at home in the numerous barracks, airfields and naval bases dotted throughout the country. With their massive luggage space, the Inter Station vehicles

C 106 joined the Railway Air Services contingent in early peace-time days, on 12th September 1945, and although the original headlights have been restored the side lamps still retain their wartime masking and the front bumper is temporarily missing. The vehicle is allocated to Victoria garage and stands at the top of the slope up from the basement there. *Omnibus Society*

Shortly after overhaul in June 1945, and with its front bumper now free of white paint and newly repolished, C 107 performs a hire duty for Channel Island Airways. Little is known about this operation, which presumably occurred in daytime hours when vehicles were no longer scheduled for Inter Station work. Channel Islands Airways' flights to Croydon Airport were renewed in June 1945 after liberation of the islands and were absorbed into British European Airways on 1st April 1947. *Alan B Cross*

were ideal for transporting scenery and other stage props, as well as performers and backstage staff from one performance to another. In this role they would have travelled widely throughout Britain but did not venture abroad. ENSA's address was given as the Theatre Royal, Covent Garden, but when not on tour the vehicles appear to have been kept at the Astoria garage in Old Kent Road, not far from their regular peace-time base. Some, and probably all, of them were painted khaki for their service with ENSA, a necessary camouflage when visiting military bases. Four of the Cubs (C 106, 109, 111 and 112) spent only a few months with ENSA and were handed back to London Transport in August 1943, but the other four remained with the organisation until the establishment of peace and did not return until the spring of 1946.

The wisdom of not selling the Inter Station Cubs, as once proposed, was justified when, on 5th November 1943, the Regional Transport Commissioner wrote to London Transport asking if the Inter Station service could be reinstated on a limited basis, primarily to cater for troops needing to travel between the various London railway termini. After consulting with the railway companies, London Transport agreed to run a service linking Waterloo, Kings Cross, Euston, Paddington and Victoria stations to meet train arrivals between 6pm and 3.55am on a schedule requiring five vehicles. No return to daytime operation was proposed either then or in the peacetime years that followed. Fortunately, the four Cubs recently returned from ENSA were available and were soon repainted back into full Inter Station livery, the fifth place in the schedule being taken by a suitably converted and blue-liveried double deck ST. The service recommenced, again from Old Kent Road garage, on 15th December 1943 and soon afterwards another ST was converted to increase the available fleet to six. In April 1944 the operation became a circular one based on Waterloo, working both ways around the circle, and in the absence of an adequate fleet of taxis passenger traffic on it rapidly increased.

Such was the passenger demand that the Cubs found it difficult to cope, and in due course three more STs were provided to cover the main service. This enabled London Transport to remove the Cubs and to transfer them, at the request of the main line railways, to operate for Railway Air Services Ltd between their London office in Lower Belgrave Street and Croydon Airport. Railway Air Services had been set up in March 1934 by the main line railway companies to operate air services internally within Great Britain, which it did until 1st February 1947 when it was absorbed into state-owned British European Airways. C 111 and C 112 were seconded to RAS in November and December 1944, followed by C 109 in April 1945 and C 106 in September of the same year, leaving the Inter Station service solely in the hands of STs. The RAS operation was covered by LT staff from Victoria garage until 22nd April 1946 when responsibility was transferred to Old Kent Road and the four Cs found themselves back in their old haunt once again.

With the war over, the Inter Station Cubs gradually drifted back to resume their old duties. Exactly the same light blue and primrose livery that proved so successful and eye-catching before the war continued unaltered, the only visible change on the Cubs being that the semaphore-type trafficators fitted from new were removed and the vacant apertures plated over. The four vehicles with ENSA were returned in April and May 1946, the last being C 107 on 13th May. This allowed a piecemeal reintroduction to begin on the Inter Station service although there were occasions when the ex-ENSA vehicles joined the four regular Railway Air Services Cubs on airport work which, during the course of 1947, became increasingly focused on Northolt Airport rather than Croydon. By June 1947 the Cubs had regained complete control of the Inter Station service and could also be found helping out on conventional bus services from Old Kent Road garage during weekday rush hours. Their airport work had ceased entirely by 9th December 1947 when the last Cub (C 109) was returned from what was now British European Airways.

Further daytime use of the otherwise largely nocturnal Inter Station Cubs found them making occasional peak hour appearances on conventional bus services operated from Old Kent Road garage on which they would no doubt have proved a great novelty to the majority of passengers. In this instance a slightly run-down looking C 108 is working as a relief on trunk service 53A. *D W K Jones*

An era ended on the night of 21st/22nd November 1950 when Cubs worked the Inter Station service for the last time. Thereafter a reduced operation was introduced using standard RTs in conventional red livery. London Transport's Press Office published an announcement on 23rd November proclaiming that "London's last petrol buses have retired, marking the end of a 46-year period", but as events subsequently transpired this was a little premature.

British European Airways was aware of their demise and in December 1950 asked London Transport about the possibility of using them for the 1951 summer season to augment the Commer Commandos which London Transport operated on its behalf between the central London terminal, which was then in Kensington, to the airports at Heathrow and Northolt. A loan was agreed on the understanding that London Transport would not have the facilities to repaint the vehicles into BEA colours, so they remained in their traditional blue and primrose. All eight were relicensed from their winter storage in Old Kent Road garage on 1st March 1951 and worked alongside the Commers from the basement of Victoria garage. Their temporary reprieve came to an end on 1st October 1951 when they were delicensed for the final time.

C 107 was the first to be sold, in July 1952, to the London Fire Brigade where it was converted into an impressive-looking emergency vehicle based at their main station on Albert Embankment. C 111 passed to the same ownership in 1953, this time for use as a personnel carrier, and in the same year the remainder were all sold to dealers. Fortunately, C 111 went for private preservation in May 1961 and has since reappeared in public in full Inter Station colours. It is thought that C 112 also still exists awaiting restoration following its earlier use as a private caravan.

Another novel albeit very occasional daytime outing for the Inter Station Cubs saw them helping out on the annual Derby Day service at Epsom Downs. On this occasion, 5th June 1948, C 111 finds itself squeezed between an ST and an STL with a little CR following on behind. *Alan B Cross*

The end is now fast approaching, but fourteen years after it was built C 109 looks almost unchanged from when it was new with the exception that its trafficators were removed in 1947 with new body panels hiding all signs of their former existence. The photograph was taken at Victoria station. *D A Jones*

The former C 110 is believed to have been the only one out of the eight Inter Station Cubs that fell into the hands of a travelling showman who retained its Inter Station paintwork. Mr A Daley of Leeds appears to be using the lower portion as a daytime living quarters and the observation saloon as a bedroom. He bought C 110 from W North of Leeds in 1955 and sold it back to them some three years later after which it was presumably scrapped. *Travel Lens Photographic*

THE HENDON TITANS – STD 1-100

During the years between its incorporation in July 1933 and the outbreak of war the London Passenger Transport Board purchased only one batch of double deck motorbuses from a supplier other than AEC. This was the one hundred strong fleet of Leyland Titans which entered service as STD 1-100 during 1937. Though minuscule in fleet size compared with the nineteen-thirties influx of more than 2,500 STLs on AEC Regent chassis, the STDs went on to outshine the STLs and to earn a reputation second to none for sturdiness and reliability, whilst aspects of their engine design influenced London Transport's engine development policy for years to come.

It is known that the senior engineering management team at Chiswick was in touch with their counterparts at Leyland right from the time the LPTB was formed. The London Transport team, led by its formidable Chief Engineer A A M Durrant, kept fully abreast of all developments within the passenger transport industry, and inevitably they would have scrutinised closely the developments at Leyland which at the time was indisputably the leader in the field of motorbus design and manufacture within the UK.

One aspect of vehicle design then very much to the forefront was the development of the

STD 92 was the first of the torque convertor-equipped vehicles to enter service, on 15th June 1937, and was photographed on the route 183 stand at Golders Green. When viewed from the offside these could not be distinguished from the main, crash gearbox batch, but on the nearside the deep slim convertor fluid reserve tank mounted on the front bulkhead was immediately apparent. *J F Higham © A B Cross*

diesel engine (or the heavy oil engine as it was generally referred to at the time), which held out the promise of considerable savings in operational cost compared to the conventional petrol unit. The experiences of the Board's engineers in this sphere, dating back to late London General Omnibus Company days and using oil engines supplied by AEC, had not been happy ones, but with their eyes constantly focused on developments elsewhere they could not have failed to notice the much better performance that Leyland was managing to achieve with its oil engines and would have wanted to study the various aspects of Leyland's design superiority for themselves.

The opportunity to do so came on 21st February 1934 when the London bus fleet of Birch Brothers was compulsorily acquired. Among the many assets taken over was a Leyland Titan TD2 which was not yet a year old and was powered by a Leyland oil engine. Numbered TD 85 in the London Transport fleet, it was one of 195 Titans inherited from a host of previous owners in 1933-35, 174 of which were bodied as double deckers. After being overhauled at Chiswick, and no doubt minutely examined in the process, TD 85 was despatched on 20th March for operation from Hanwell garage alongside the rest of London Transport's then small oil-engined fleet, Hanwell being the location at which experimental vehicles were traditionally located in pre-war years. Its engine was Leyland's standard overhead camshaft 8.1 litre six-cylinder direct injection 87bhp unit, rated at 46.8hp with bore and stroke of 4⅜ins x 5½ins and featuring aluminium alloy pistons with toroidal formation in the piston crowns. It must have impressed its new owners who lost no time in placing an order for four more of the same, which were received in October 1934 and immediately placed into further recently-acquired Titans which happened to be going through overhaul at the time. These were TD 94, 95 (previously in the Westminster fleet), TD 98 (ex-St George) and TD 105 (ex-Miller).

All four had joined TD 85 in service at Hanwell by November 1934 during which month, on the 7th, the City Motor Omnibus Company was taken over, bringing four more Leyland oilers into London Transport ownership. One was a Titan, TD 121, and the other three were TT1-type Titanic six wheelers TC 1-3. All were powered by the same type of oil engine with which London Transport was now becoming familiar, and they duly joined the other oilers at Hanwell. Subsequently the fleet of oil-engined Titans was increased to nine when the three Titanics were converted to petrol and their engines were placed into TD 96, 124, 130 at overhaul between December 1935 and April 1936.

At about this time, London Transport signified to Leyland its intention of purchasing 100 brand new Titans fitted with the manufacturer's own metal-framed bodywork. The exact time and circumstances of this notification are not known, but it must have been early in 1936, for on the 19th March of that year newly-built STL 1217 was despatched to Leyland's works in Lancashire for it to be thoroughly studied and measured in order to understand and replicate London Transport's bodywork requirements. STL 1217 returned to Chiswick nearly seven weeks later, on 28th April. It was not until 17th August 1936 that a Special Expenditure Requisition was approved giving permission for the order to be officially placed at a total outlay of £182,540 with a completion date for delivery of all 100 vehicles by May 1937 at the latest.

London Transport's purpose in obtaining the STDs was twofold. It wished to test the alleged superiority of the Titan – which was so frequently proclaimed by other operators – against the AEC Regent which it was obliged to purchase, whilst at the same time establishing whether it was, in fact, getting value for money from AEC. In particular, Durrant and his team were far from convinced that AEC's 7.7 litre oil engine was as efficient as it should be, and they strongly suspected that the indirect fuel injection system to which AEC was wedded was less efficient than the direct injection method employed by Leyland, causing operating costs to be higher than they needed to be. The acquisition of a fleet of Leylands of meaningful size would give them the opportunity to put this theory to the test.

In order to ensure a fair comparison between the two makes it was necessary for London Transport to accept as many standard Leyland features as possible on the new chassis and to resist the temptation to incorporate many of its own standard components. Therefore, in placing the order, London Transport stipulated basic Leyland-designed units such as engine, clutch, gearbox, brakes, axles and differentials. The Titan chassis differed from the Regent in that it incorporated no built-in extension to support the platform and staircase but used instead a bolt-on section at the rear, and this too was accepted by London Transport. However, the Board's engineers could not resist specifying a few modifications from the standard TD4 design, which Leyland were happy to go along with to clinch such a large and prestigious order. Thus the chassis frame itself was modified, notably by having a wheelbase of 16ft 3ins which brought it in line with London Transport's STL fleet but was 3ins shorter than the standard Titan design. The frame was modified at the front to incorporate a revised type of spring anchorage, also making it compatible for towing purposes with the STL. The result was that the dumb irons were square-shaped instead of rounded, a feature subsequently adopted by Leyland itself as a standard fitting with the introduction of their TD5 model in the autumn of 1937. Worm and nut type steering gear supplied by AEC was fitted in place of Leyland's standard Marles cam and roller system. On the other hand, Leyland persuaded London Transport to experiment with RP-type brake adjusters in place of its own favoured type based on the Clayton-Lockheed vacuum braking system, initially as an experiment on the first six chassis, but this quickly proved so successful that it was used finally on all one hundred. As delivered, the STDs were fitted with 36x8 tyres at the front and slightly smaller 900x20 ones at the rear.

At the heart of the STD was the latest version of Leyland's standard oil engine, designated E39/2. This was a slightly more powerful 8.6 litre variant on the original design, which it closely resembled in all ways including the use of pot cavity combustion chambers, but now with a slightly enhanced bore of 4½ins giving a rating of 98bhp at 1,900rpm. It was not, in fact, completely new to London Transport as one of its original Titan oilers, TD 94, had been sent to Leyland in April 1936 to be fitted with one such engine. As was usual London Transport practice, the engines for the new STDs were set to run in slightly de-rated form at a maximum of 1,850rpm. London Transport was already aware that this was a notably fuel efficient engine which, through being of the direct injection type, avoided the wasteful oil consumption and relatively high piston ring failure inherent in the indirect injection system. There was also an expectation that a big reduction in starter troubles would be recorded. Probably the best and most highly respected of all 6-cylinder oil engines on the market in pre-war years, the Leyland 8.6 was notable for its quiet, smooth running, while its responsiveness in all traffic conditions made the Titan (and its single deck counterpart, the Leyland Tiger) a sheer pleasure to drive.

The first ninety vehicles (STD 1-90) were specified with Leyland's standard transmission system which included a single dry plate clutch with clutch stop, a conventional crash-type gearbox with constant mesh 'silent' third speed wheel mounted on needle bearings, a fully-floating rear axle with 7½ins worm centre, and with the manufacturer's standard ratio of 5.4:1. It may have seemed odd for London Transport to specify crash gearboxes when it had become wedded to preselectors and fluid flywheels for its large STL fleet, but at the time all of its drivers were well versed in the use of crash gearboxes and, as time would tell, the STDs appeared to suffer no disadvantage whatever through having fully manual transmissions, even in the most demanding of traffic conditions. The low-pitched and sometimes very loud roar from the gearbox when pulling away in 2nd from a stop or from traffic lights, and the subdued but quite musical tone in 3rd gear, were unmistakable characteristics by which the vehicles could always be easily recognised, while in top gear they bowled along quietly, achieving quite high speeds with seeming effortlessness.

The final ten vehicles of the batch (STD 91-100) differed from the remainder in not having conventional gearboxes. Classified by Leyland as type TD4c, they were specified with HTC3-type hydraulic torque convertors of the Lysholm Smith type promoted by Leyland as an optional extra, which put a premium of £82 on the purchase price of each chassis and added 189lbs in extra weight. This order marked the second stage of an experiment being carried out by London Transport with the aim of enhancing the ease of driving beyond the advantages of the preselector gearbox to a system of 'gearless' two-pedal control, virtually eliminating the need for most of the skills which conventional gear-changing demanded.

Operation of the system was simple. The driver was provided with a control lever mounted at floor level in about the same position as a conventional gear lever. Pushed into the forward position, known as convertor drive, the engine was coupled to a pump supplying oil to a three-stage turbine linked to the propeller shaft through a free-wheel. The velocity of oil from the pump was dependent upon the engine speed and the torque depended on the impact force on the turbine blade driving the propeller shaft. The driver would start the vehicle from rest with the control lever in the convertor drive position, and when the bus achieved somewhere in the region of 15mph the lever was moved to the rearmost and almost vertical position known as direct drive to couple the engine direct to the transmission, the turbine being disconnected on account of the free-wheel. The vehicle could be brought to a halt with the lever still in the direct drive position. By moving the lever to the left when stationary the driver could engage safety neutral, which was essential when starting the engine to prevent accidents caused by forward creep, and further left, beyond this, was reverse. Needless to say, efficient operation of the system demanded an always-adequate supply of convertor fluid which consisted principally of high boiling point kerosene free from dissolved gas, with 5% of engine oil added for lubrication purposes. For efficient operation, it was essential to ensure that the fluid was always perfectly clean, and a header tank, which always had to be fully filled, was prominently mounted on the front bulkhead and was an unmistakable sign that the vehicle was fitted with a torque convertor.

In the Leyland body shop two rows of newly-built STDs receive their final touches with STD 10 appearing to be virtually complete. The paintwork and sign-writing has all been done and, in compliance with normal London Transport practice, the main large advertisement panels on each side have been left unvarnished. It was reported at the time that Leyland Motors took on additional manufacturing staff to fulfill this large contract. *British Commercial Vehicle Museum*

London Transport's interest in the Leyland torque convertor became apparent with a November 1933 proposal from A A M Durrant that a single unit should be obtained from Leyland on suitable terms for fitting into an AEC Regent subject to the concurrence of the Managing Director of AEC. The intention at this time was that two further torque convertors of the same type should be obtained for an extended trial if the first one proved satisfactory. STL 221 was selected for the experiment, and it ran from Clay Hall garage with a torque convertor fitted until August 1935 when it was transferred to Upton Park to join Leyland Titan TD 83, which on the 23rd of the same month had also been fitted at Leyland with a torque convertor, perhaps a slightly updated version of the one in the STL. TD 83 only retained its convertor until 23rd November 1935 but STL 221 was still running in this form when the STDs were ordered in 1936, and the torque convertor was not removed from it until August 1937.

The body contract for the STDs specified that they should comply with the manufacturer's standard all-metal structural design, retaining all standard profiles but with detailed finish amended where appropriate to accommodate London Transport's wish to achieve similarity in styling to the STL. This was an important contract for the Leyland body shop which had earlier been threatened with the loss of an order for a similar number of trolleybus bodies for London Transport following the revelation that very serious structural problems were being encountered by many other purchasers of Leyland all-metal bodies. As a result of these problems, which ended up costing the manufacturer dearly, a completely new management team had been recruited by Leyland to head its bodywork design and construction functions, and they had rapidly set about producing a totally new design and getting it into production. London Transport had specified that a specimen body to this new design should be made available by 16th December 1935 if the order for 100 trolleybus bodies was to still go ahead, but Leyland could not meet this date and the order for 99 of them had gone instead to Metro Cammell. However, a single trolleybus (no. 384) was eventually taken into stock on 2nd April 1936, almost five months late, and found to be satisfactory, and it was probably on the strength of this that the order for 100 STD bodies was given the go-ahead.

Fortunately, Leyland's new double deck body design, which was attractively simple and unlikely to date quickly, was very similar in profile to that of the STL. This meant that once London Transport's

STD 17 was received at Chiswick on 27th April 1937, and after all the advertising material had been applied it was fully fitted out with blinds and route number plates ready for the official photographs to be taken. The general similarity in shape of the standard Leyland body shell to the contemporary STL is immediately obvious, and the most unsatisfactory design mismatch is probably where the deep front canopy required by London Transport conspicuously fails to match up with the contour of the driver's windscreen and the panel above it. Although initially painted silver, the surrounds to the roof number box were changed to red before entering service. *London Transport Museum*

styling requirements had been applied the STL and STD did not look so very different from each other, and a close family likeness was certainly evident. Leyland's new standard steel frame design was specified and London Transport required that, where timber was used, it was to consist of American ash for fillets and packings and English ash for any structural work. Built to a length of 25ft 11ins, the STDs achieved a fairly low overall height of 14ft 4¾ins partly thanks to the fitment of springs which were as flat as possible to minimise the overall height. The standard seating capacity of 56 was achieved on STD 1-90 (30 upstairs and 26 down), but on the gearless vehicles this had to be reduced to 55 by fitting a single seat instead of a double on the upper deck offside immediately ahead of the staircase. This reduction was necessitated by legal maximum weight requirements to offset the greater heaviness of the chassis, giving an unladen weight per seated passenger of 267lbs on STD 1-90 and 275lbs on STD 91-100. Both types were recorded at 10.6 tons fully laden with five standing passengers, and their officially quoted unladen weights were 6tons 13cwt 3qrs and 6tons 15cwt 0qrs respectively.

Since borrowing STL 1217, London Transport's standard design had been updated to include a route number box within the front dome, and the STDs were built with this feature in place. Viewed externally, the STD generally resembled a slightly pugnacious version of the roof-box STL thanks to the higher-mounted and more aggressive-looking radiator and sharper curve at the foot of the driver's windscreen. While the cab front followed Leyland's standard profile, the offside of it incorporated London Transport's standard 1ft 4½ins open doorway separated from the bulkhead by a 10ins wide glazed partition, an arrangement which must have surprised Leyland's coachbuilding staff accustomed to fitting drivers' doors as a matter of routine. To the nearside of the cab, London Transport specified a deep canopy, a feature which Leyland did not normally provide and which did not sit comfortably relative to the profile of the windscreen and the ventilator above it (another London Transport requirement), projecting slightly forward of them. At the sides and rear, Leyland's normal curved waistrail panel was omitted in favour of London Transport's standard arrangement, and a London-style conductor's signalling window was provided on the offside although it was only about half the depth of the equivalent fitment on the STL.

A typical offside view of an STD, in this case STD 38 which began work at Hendon on 19th May 1937. The same open-doored driver's cab as on the STL with its small secondary window was a definite break from the standard Leyland body design while, towards the rear, the route number stencil holder was mounted higher than on the STL to line up with the much smaller conductor's signalling window. *J F Higham / A B Cross*

At the rear, Leyland's standard-profile dome incorporated a typical London Transport Y-style emergency exit which was, of necessity, shallower than on the STLs, and all around the centre of the vehicle Leyland's characteristic narrow, scrolled beading was used to separate the two decks rather than the deeper, flat moulding usually favoured by London Transport. Other points of difference from the STL included a straight lower edge to the foot of the front bulkhead windows at a height dictated by the bonnet line, the omission of rounded corners from the tops of the upper deck front windows, and the fitment of a 'Simplastic'-glazed window with radiused corners on the platform bulkhead and on the offside of the driver's cab. With the exception of the front nearside one all the mudguards were rubber, and access to the 35-gallon fuel tank was on the offside. All the external panelling was in aluminium except for the lower back panel and corner which were steel. The livery application was exactly the same as on Central Bus STLs except on the rear dome where the silver of the roof was continued right down to the bottom line of the windows, although the emergency exit itself was picked out in off-white.

London Transport specified, when issuing the contract for the 100 bodies, that "the interior appearance in all respects is to be equal to, or superior, to that of the Board's standard 56-seater double deck bus of the STL11 type", and they were. Their seating layout was exactly like that of the STLs, including the unusual pair of single seats on the upper deck nearside adjacent to the stair well, and other standard fittings included 'Novobac' ventilators at the front of each saloon, and two long, ash ceiling rails on the upper deck. Many of the internal fittings were, in fact, supplied by London Transport including all the lamp and bell equipment. The staircase and platform also followed standard STL arrangements as did the tubular framed seats, cushions and coverings. At the front of the lower deck it was specified that the word 'Leyland' should not appear on the flywheel cowl, and on STD 91-100 an additional floor trap was provided to give access to the torque convertor radiator vent plug. All the interior mouldings and side panels were covered in rexine, with the lower halves of the window frames in brown and the upper parts in pale yellow, offset by a pale green band above the windows and green on the side panels and seat backs.

Compared with the colourful interior, the driver's cab was very spartan in the usual London Transport tradition, and it was specified that visibility from it should be at least as good as that on the STL. No instrument board or speedometer was provided, but London Transport supplied its usual emergency saw, strap and saddle for fitting on the bulkhead above the driver's head and a Pyrene fire extinguisher was hung on the bulkhead to the offside of the driver's seat.

The contract for the STDs specified that the first vehicle should be delivered to Chiswick in the week ended 30th January 1937, followed by a gentle build-up through February to a maximum of 10 per week from 6th March onwards with the final delivery taking place in the week ended 1st May. In fact, the programme took longer to get under way than anticipated and the first complete vehicle, STD 91, was not received at Chiswick until 18th March 1937. It went through extensive tests there on 1st and 2nd April in unladen, half laden and fully laden conditions, where its performance was recorded in detail. The operation of the torque convertor was of particular interest to the engineers and was closely monitored. STD 91 was returned to Leyland on 20th May for adjustments to be made to the torque convertor, not returning until 28th June during which month most of the remaining torque converter vehicles were also received. Meanwhile delivery of the crash box vehicles began with STDs 1 and 2 on 5th April 1937, and the final STD delivery, STD 100, took place on 3rd July. Similar tests to those on STD 91 were carried out on crash box STD 87 on 16th to 18th June enabling preliminary comparisons to be made regarding the performance of the two types. These showed the torque convertor vehicle to be the more lively of the two, attaining 15mph in 6.8 seconds as against 10.0 seconds on STD 87, and 30 mph in 35.3 seconds (torque convertor) compared to 44.5 seconds (crash gearbox), the latter figures being based on a load equivalent to half the seats being filled. On the other hand, when it came to fuel consumption tests, STD 87 proved consistently better to the extent of about 0.4mpg.

With the one obvious exception that the bottom edges of the front bulkhead windows did not slope down to their outer corners, the lower saloon interior was almost exactly identical to that of the STL, including the provision of two 'Novobac' ventilators in the front bulkhead which could be locked open or shut as required, and the three advertisement panels below the front windows. Leyland's staff had obviously studied the interior fixtures and fittings of STL 1217 in great detail when it was on loan to them with the aim of mimicking it as closely as possible. *London Transport Museum*

In a similar manner to the lower deck, the overall appearance in the upper saloon differed little from that of the STL. This view taken in the factory looking towards the rear of a newly completed STD shows that the standard London Transport colour décor and green seat upholstery are in evidence as is the unusual seating arrangement at the rear with two single seats adjacent to the stairwell giving access to one further double seat beyond. *British Commercial Vehicle Museum*

September 1937, and the fleet at Hendon garage now consists one hundred percent of new STDs. As a crew changeover takes place in Station Road, West Hendon, two of the class meet up with STD 64 in front covering a Sunday working on route 83A. The vehicle behind, on route 183, is clearly one of the torque convertor batch but its fleet number cannot be ascertained. *G H F Atkins*

STD 1-100 carried chassis numbers TD4/12007-TD4/12106 which were in numerical sequence as were their London Transport body numbers 17335-17434, and these bodies all stayed with their original chassis throughout their time with LT. Registration numbers were also in a numerical block DLU 311-410 and the vehicles were classified as 1STD1 (STD 1-90) and 2STD1/1 (STD 91-100).

Most unusually for London Transport, the STDs were ordered with no pre-conceived idea in mind as to where they would be used, except that it would be somewhere in the Central Bus fleet. The first idea, propounded in September 1936, was that they should be allocated to Putney Bridge garage although, as this was not quite large enough to accept them all, two would need to be based at nearby Chelverton Road. This proposal was quickly ruled out and Hanwell was mooted a month later as an alternative, where the STDs would work alongside nine oil-engined early Titans, 11 Gardner-engined 'Bluebird' LTs and five standard LTs. Frank Pick was not entirely happy with this and suggested that A A M Durrant might consider purchasing five more STDs to replace the five standard LTs at Hanwell. Finally, it was decided that Hendon would be the base from which the STDs would operate, and this turned out to be a happy decision. The vehicles were well looked after at Hendon and the staff in the garage, both operating and engineering, held them in high regard. Hendon and the STDs became synonymous for the next decade and a half.

At the time Hendon's scheduled Monday to Friday run-out was 89 vehicles on no fewer than seven routes: 13 (24 NS), 28 (20 NS), 52 (12 STL), 83 (10 ST), 113 (6 STL), 121 (7 ST), 183 (10 ST). As routes 13 and 113 traditionally had no Sunday operation, Hendon always had to be found additional work elsewhere on this day of the week which, at the time, was on route 16. A year later, on 1st June 1938, an increase in Monday to Friday work on route 28 brought Hendon's scheduled allocation to 95, which was just right to accommodate the 100-strong fleet. A single vehicle, STD 6, was sent to Hendon on 21st April 1937 and was the first oil-engined bus ever to be allocated there. It may not have started work on this date, but probably waited until the 23rd and 24th when sufficient additional vehicles joined it to enable the full conversion of route 52 to take place.

New STDs were now pouring in and were stockpiled at Chiswick tram depot and in Leyland's works at Kingston ready for their mass introduction on 17th May on all of Hendon's ST and STL-worked routes. Replacement of the NSs on the 13 and 28 completed the process between 31st May and 2nd June. At this stage none of the torque convertor vehicles had been put into service: the start of this did not occur (with STD 92) until 15th June. With the exception of STD 90, which for some unknown reason was a very late arrival (on 15th October), the whole class was in use by 13th July 1937.

Once the STDs had settled down in service at Hendon, London Transport's technical boffins began a full-scale examination of their operating performance, using 89 STLs of the same age at Battersea garage as the basis for comparison. The road failures, and the reasons for them, for both groups of vehicle were recorded and analysed over a 15-week period from 15th September to 21st December 1937. A report was subsequently produced which summarised the results thus: "The performance of the Leyland vehicles has been eminently satisfactory, and, in fact, surpasses that which has been obtained from the A.E.C. vehicles, particularly in relation to the performance of the engine and the brakes". This was, in fact, an understatement. In the case of engines alone, five had to be changed on STDs during the control period compared with 26 on the STLs, while the total number of road calls for other reasons (excluding punctures) amounted to 64 on 100 STDs as against 103 on 89 STLs.

Additionally, the STDs were proving to be 4% more efficient in terms of fuel consumption. These results were put to management in June 1938 but were met, in some quarters, with disbelief. Therefore, it was decided that a further test should be organised after support staff from Leyland, who had been working alongside the engineers at Hendon garage, had been removed. The Leyland staff departed in September 1938 and a further trial was undertaken over 13 weeks between 27th September and 30th December 1938.

Photographed outside Burnt Oak station, STD 88 has just been put through its first overhaul, in March 1939. A minor livery variation visible on its new coat of paint is that the driver's cab window frames are no longer picked out in back, and a small but visible modification to the bodywork is the new flat-profiled drip moulding above the driver's cab door. STDs were then nearing the end of their days on route 52 subsequent upon the opening of Alperton garage on 7th June. *Charles F Klapper © Omnibus Society*

The result of the new test, published in June 1939, was broadly similar to the earlier one and concluded that the STDs were, in effect, an incredible 24% more efficient overall than the STLs. In terms of operated miles per failure for all mechanical reasons, the STDs at 7,100 showed a massive improvement over the STLs at 5,100. However, long before this report was produced, the greater reliability of the Leyland engine, particularly in regard to piston and bearing failures, had already been fully acknowledged by London Transport's engineering management, whose total belief in the superiority of direct over indirect injection had also been fully vindicated. In fact, they had already confronted AEC with their findings, and on the assertion that current AEC engines were, in effect, involving London Transport in much unnecessary expense, they had instructed AEC to produce a direct injection engine equal to the Leyland in operability, power and reliability for their forthcoming class of 10T10 single deckers.

The furthest point south to which STDs were ever scheduled to operate was Crystal Palace on the 1939 Sundays-only route 3A, where STD 40 is see at the time-honoured bus stand in Crystal Palace Parade on 7th May. The 3A ran for the last time on 15th October 1939 and the most southerly destination that Hendon's STDs ever regularly reached thereafter was London Bridge. *Capital Transport Collection*

The only downside, from a Leyland point of view, of the latest trials was that the ten torque convertor vehicles were found to be producing unsatisfactory mpg results while the convertors themselves were proving to be mechanically unreliable in a number of ways.

One of the reasons for purchasing the STDs, which was to compare their performance with the AEC product as a means of judging the efficiency of the latter, was now concluded. With regard to whether the AEC Regent gave good value for money, the Board's auditors found themselves unable to come to a definite conclusion. The STD fleet had cost a total of £181,705, which was an underspend of £835 on the authorised amount. A true comparison of chassis costs proved impossible because no equivalent number of AEC Regents had been purchased in recent times to mitigate the effects on price of bulk buying, all STL orders having been for far greater quantities. The fact that substantial cost savings could be negotiated through buying in quantity is illustrated by the fact that the official list price for a single Leyland TD 4 chassis was £1,000 whereas, when buying a batch of 100. London Transport had been able to negotiate a payment of £829 per chassis (or £907 with torque convertor), and this was after accepting numerous differences from the standard product which would have added to their cost. Inevitably, the savings resulting from bulk buying of AEC Regent chassis in far greater quantities would have been even higher. The auditors' final summary said that "while the average cost of the Leyland chassis bore a reasonable relation to that of the Associated Equipment Company chassis there was no conclusive evidence as to which make was the cheaper". Body costs were not comparable because the Leyland bodies (at £980 apiece) had a main structure of steel whereas Chiswick-built bodies were of a cheaper construction of timber reinforced with steel flitches. The Leyland bodies were, however, undoubtedly a better bargain than 175 all-metal bodies bought in 1937 from Park Royal on STL chassis which cost £1,006 each and – as London Transport was to find to its detriment a decade later – were of decidedly inferior construction.

The first overhaul cycle for the STDs commenced when STD 6 was taken into Chiswick on 31st October 1938 and continued until the last, STD 90, appeared newly overhauled on 15th June 1939. In a small livery change, the white window surrounds on the upper deck now continued right around the back of the vehicle, as on STLs and other double deckers in the fleet, instead of the silver roof colour being extended down to upper waistrail level on the rear dome. The second change came about as a result of complaints from crews and passengers about rain falling from the bodywork and splashing on to them. London Transport decided to fit what it called 'drip mouldings' above the driver's doorway and over the passenger entrance in place of Leyland's standard rounded moulding. These 'drip mouldings' consisted, in fact, of standard flat-profile, raised sections as used on the STLs and other classes, and inevitably they did not match up with the original Leyland fittings on either side of them. Because both were painted black, this difference was barely noticeable until, many years later, the livery style changed and these mouldings were picked out in cream. A third modification, incorporated at drivers' request, was a push-out ventilator in the front cab panel clearly visible just above the offside headlight.

Although they had only been in use for two years, the end came rapidly for the torque convertors in 1939. No evidence exists to say that drivers disliked them, but the engineers certainly did! They had been inundated with problems, particularly in relation to leaking glands, defective free-wheels and excessive wear on the clutches within the convertors. The units needed constant maintenance to keep them in good order, and weekly checking and cleaning was required on all the injectors and reserve tank filters. Absolute cleanliness was essential – which was not always easy to ensure within a working garage environment – as seals within the convertor were prone to failure through lack of cleanliness or because of an incorrect or poor quality mixture of kerosene and engine oil. In a single year no fewer than 24 torque convertor failures occurred in a fleet of ten vehicles. Further to all this, there was the poor mileage per gallon inherent in the system to be taken into account. The fact that failures and low mileages between

A wartime winter scene inside Cricklewood garage finds a host of buses overnighting with their radiator covers in place as protection against the cold. Most are STLs plus a sprinkling of STs, but fourth from the right, distinguished by the unmistakable line of its cab window, is STD 81. The third vehicle beyond this is also an STD, and both carry blinds for route 16 which appears to be the only Cricklewood route upon which the Leylands ever operated. *London Transport Museum*

breakdowns continued to be obtained even after units had been overhauled by Leyland led to the decision that the torque convertors should be removed and replaced by conventional crash gearboxes.

It was arranged with Leyland that the vehicles would be sent to their works at Kingston at the rate of two per week starting with STDs 93 and 98 on 19th June 1939. Initially only eight out of the ten were scheduled to be dealt with as STDs 97 and 99 had been experimentally fitted with a modified torque convertor, and it was agreed with Leyland that these two should stay in service for a little longer to accumulate operating experience with them. This modified convertor contained an alteration to the blade angle, which it was hoped would produce a marked improvement in acceleration and hill climbing, and also improved seals which theoretically required less attention. The last vehicle under the original programme, STD 95, was despatched to Leyland's on 8th July, leaving STDs 97 and 99 to operate into the early months of wartime before the last one of all, STD 97, received its crash gearbox on 13th December 1939. After the conversion work had been carried out on each individual bus, it returned to Chiswick to be converted to a 56-seater and was recoded from 2STD1/1 to 1STD1.

In the meantime, on 7th June 1939, the new garage at Alperton had opened for business, and this impacted heavily upon Hendon and its STDs. From a scheduled run-out of 95 buses, its Monday to Friday output plummeted to only 58. The whole of its allocation on routes 28, 52 and 83 (plus 83A Sundays) was lost, and the only compensation for this was the receipt of four workings on the double deck section (Golders Green–Edgware) of route 240. The Sunday allocation on route 16 was also lost, although this was compensated between May and October 1939 by some work on routes 3A and 112. Oddly, no STDs were taken out of service or transferred elsewhere for operation, leaving Hendon with an enormous, unused surplus. This probably indicated that future plans were afoot to build the garage's work up by some other means, but war intervened.

Heavy troop movement activity just prior to the outbreak of war led to STDs 69 and 90 being sent to report at the drill hall in Lemsford Lane, Welwyn Garden City on 24th August 1939. STD 69's movements under military control took it no further than Biggin Hill but STD 90 travelled to Suffolk, firstly to Martlesham near Ipswich and then to Felixstowe, which would have been one of very few occasions when a journey was made by any of the STD 1-100 batch beyond the London Traffic Area boundary while in London Transport's ownership.

The sudden cessation of all Green Line operation and the implementation of extensive evacuation plans on 1st September 1939 – two days before war was officially declared – impacted immediately upon the stockpile of unused STDs. Thirteen immediately left Hendon for temporary service at three Country area garages to assist with evacuation and troop movements, and also to cover normal scheduled operations if needed. Hatfield received STD 15, 22, 56, 58, 74, Luton STD 29, 39, 47, 68 76, and Staines STD 42, 81, 95. The five at Luton were subsequently transferred, on 12th and 13th September, to St Albans. By the end of September they were no longer required by the Country Bus & Coach department and all thirteen were delicensed on 30th September. Even after having lost these vehicles, Hendon was still left with a surplus of 20 STDs for which no work was available and these, too, were delicensed on 30th September. One third of the class was now in storage, and for the record the 33 vehicles out of use were STD 1-20, 22, 29, 30, 32, 39, 42, 47, 56, 58, 68, 76, 81, 95.

Meanwhile all the STDs, whether in use or not, were rapidly equipped with lamp masks and white blackout markings in common with all other road vehicles, and their silver roofs were overpainted in grey as a camouflage measure, although this was replaced by bauxite (a chocolate brown colour) from the latter part of 1941 onwards.

It was clearly an untenable situation to have so many excellent and modern vehicles out of service, especially as they were so much more economical to operate than the many petrol-engined double deckers still in use. Work had to be found for them, and on 1st November 1939 all the delicensed STDs were put back into service. A few returned to Hendon, but 25 found a new home at Cricklewood garage, officially on route 16. This was a sensible arrangement as the two garages were within the same engineering group and the fitters at Cricklewood, which was the 'parent shed' for the group, were already well versed in handling STDs which went there regularly for major docking. The influx of STDs at Cricklewood released other vehicles, some of which were used for the double decking of route 231 at Chalk Farm.

The number of STDs at Cricklewood fluctuated in line with changing vehicle requirements at Hendon. The latter garage was faced with the loss of route 121, which was amongst those withdrawn on 18th October 1939 as a wartime economy measure, but partial compensation was provided by the extension of route 113 southwards from Hendon Central to Oxford Circus. Similarly, Sunday route 3A became a wartime casualty after its last day of operation on 15th October, but the introduction of a Sunday operation on route 113 partly made up for its loss. From 22nd November 1939 Hendon gained a Sunday presence on route 142 and. after losing its small route 240 allocation on 1st December, the garage and its STDs settled down to a pattern of operation that lasted for many years, right through to the end of the STD era and beyond, of 13 (Mon-Sat), 113 (daily) 142 (Sundays) and 183 (daily).

Over time, schedule requirements at Hendon built up to meet changes in demand to the extent that a Monday to Friday requirement for 73 STDs at the 13th May 1940 summer schedule rose to 87 at the start of the winter programme on 20th October 1942. By now the STDs had been fitted with restricted blind displays over a period of months in 1941/42, while subsequently, between December 1942 and March 1943, a small number – probably six – received the wartime livery of brown and cream upon overhaul at Chiswick tram depot. The number of STDs at Cricklewood had meanwhile dwindled to 12, and on 28th October these all departed, half of them returning to Hendon and the other six to pastures new at Victoria.

Victoria was the fleet's newest garage, having opened as recently as 20th March 1940. At the start it had a standardised fleet of STLs, but 1941 witnessed diversification with the arrival of a few RTs in October and the first of a new batch of Leyland Titans, STD 101, on 1st December. The remainder of this batch (STD 102-111), which carried London's first 'utility' style bodies, followed between May and August 1942. It was a logical move for the pre-war Titans not required at Hendon to join them, since the two batches had many mechanical units in common. STD 102-111 were joined initially at Victoria by STD 22, 24, 47, 49, 56, 58, but further vehicles subsequently arrived from Hendon in 1943 and 1944 bringing the maximum in stock at any one time to 12.

Operationally, the new utility STDs turned out to be a disaster, being grossly unpopular with drivers and operationally very challenging to keep to time with. Totally the opposite, in fact, to the Hendon STDs. They were built on the TD7 chassis introduced in 1939, and this design employed a system of flexible engine mountings aimed at giving smooth running but necessitating a larger and heavier flywheel to provide steady idling, with the result that gear-changing was painfully slow. Coupled with high gearing, this made the vehicles difficult to drive and largely unsuited to the heavy in-town work to which London Transport allocated them. *(More details about STD 101-111 can be found in 'London's Utility Buses' published in 1997.)*

The two groups of STDs were not officially scheduled to work together at Victoria. The utilities were intended for route 10 and the ex-Hendon vehicles for route 16, but the staff at Victoria were notably more relaxed about keeping vehicles to their scheduled routes than was the case at many other Central Bus garages, and the ex-Hendon STDs were noted to have strayed on to routes 22 and 52, and probably on to other Victoria workings too. From October 1944 their official role on route 16 had been usurped by STs and they were then allocated to route 22. The fact that they remained at Victoria into the start of the post-war era was a little surprising in view of the fact that Hendon's scheduled vehicle requirement had risen to 96 by October 1944 (exactly equivalent to its maximum pre-war level), which theoretically called for the whole of the STD 1-100 batch being allocated 'back home'. In fact, the deficit was made up by allocating STs to Hendon until 11th July 1945 when, with the receipt of new G class Guy Arabs, Victoria finally relinquished the last eight of its Hendon STDs.

A minor wartime bodywork modification approved in December 1942 was the fitting of draught prevention fillets to the cab door apertures on some 2,500 buses at a modest overall cost of about £400. The huge fleet of STLs was the main beneficiary but the STDs were also included. The so-called fillet consisted of a vertical strip of wood protruding from the body side just ahead of the driver's cab entrance which, though only a simple and barely noticeable addition, was very effective in reducing the severity of draughts caused by the lack of a driver's door. Apart from the provision,

STD 49 was absent from Hendon for most of the war. Transferred to Cricklewood in December 1939, it subsequently passed to Victoria and did not return to its home garage until July 1945. Now carrying a restricted blind display and equipped with a full set of window mesh, it is seen at the Putney Heath terminus of route 22 while working from Victoria garage. *W J Haynes*

in 1943, of boxes in the bus cabs for the storage of cash takings to counteract a rise in robberies from conductors, particularly during blackout hours, no other modifications were made to the STDs during the war, from which all one hundred vehicles emerged virtually unscathed.

The early post-war years marked a period of stability for STD 1-100. Despite the enforced lack of full maintenance caused by the unavailability of adequate manpower and materials, they had survived the war in remarkably good condition and required none of the extensive body overhauling and reconstruction that all the other pre-war double deck classes now needed. The soundness of their original design and construction shone out. An overhaul programme that had commenced during the war was completed in July 1946 and another commenced in July 1947. No body exchanges were carried out during these overhauls and each individual STD chassis kept its own body throughout its whole working life. It was probably soon after the war that the unladen weight on all of them was amended to 6tons 15cwt, although it is doubtful if any physical alteration occurred to justify what was probably merely a paper exercise to show a more accurate weight than before. Meanwhile, back in May 1946, London Transport's policy of blanking-out all disused areas of destination screens in black was changed to red on the grounds of improved appearance. Although this was intended to be applied only to vehicles when overhauled or repainted, a few garages – of which Hendon was one – took it as an instruction to treat all their vehicles immediately, so within a few weeks the appearance of the STDs was subtly changed as red paint replaced black on the indicator box glasses.

The first vehicle to emerge from the works at the start of the penultimate overhaul cycle was STD 6 on 29th July 1947, and the big surprise was that, unlike other double deckers in the fleet with a modern body profile such as the RTs, STLs and even the newly-delivered STDs (STD 112-176), they were not repainted into the new red and cream livery that had started to become commonplace since its introduction in May 1946. For some unknown reason, STD 1-100 were destined to continue carrying the old red and white styling, with black above-window mouldings and brown roof, and this continued to be the case until February 1949 when a trial vehicle was repainted upon overhaul into the modern livery. Such was the London Transport bureaucracy at the time that it required a full meeting of the Main Technical Committee on 29th March 1949 to decide that this could now be adopted as standard for this batch of STDs. During this overhaul cycle most of the vehicles dealt with up to December 1948 had their now-disused offside route number stencil holder removed, a process which ceased in 1949 when a half-hearted attempt was made to bring them back into use and some vehicles had a stencil holder put back in again.

In the months immediately following the end of the war Hendon's STDs exuded an air of wellbeing and solid reliability distinguishing them from many other double deckers in the fleet, and their deep, resounding tone echoed reassuringly against the high buildings lining the busy streets of central London. They were, in fact, the only pre-war double deckers to emerge from the war not requiring extensive body renovations. Suitably refreshed after an overhaul in September 1945, STD 80 pauses at the stop at Aldwych with a Bluebird LT close behind while on its way to London Bridge. *Omnibus Society*

A rear end view of a pair of STDs at London Bridge station with STD 36 nearest the camera shows the flat drip moulding fitted above the platform entrance on all one hundred vehicles a decade earlier. STD 36 displays the original wartime style of restricted destination blind while the vehicle in front carries the post-war version. On both vehicles the back platform window has been divided into two, a modification carried out during the war years and just after on much of the pre-war double deck fleet. *D Okill*

On most vehicles overhauled from 1947 onwards – both buses and trolleybuses – the offside route number stencil holders were removed as typified by STD 1 which emerged from overhaul on 20th January 1948. It is seen at the Oxford Circus terminus of route 113 six months later in the company of LT 283 on route 12. This was a time of great change on the London bus scene and rapid inroads were being made into the ranks of the LTs although they continued to hold sway on route 12 until May 1949. *Alan B Cross*

The only one of Hendon's main operations where their STDs were regularly pitted alongside post-war vehicles was route 13 for which Cricklewood garage received RTs in February 1948. By 1951, however, Hendon's crews were becoming insistent that they, too, deserved some new vehicles. RT 372 and STD 5 represent the two types on the terminal stand at Golders Green. *D A Thompson*

The STDs began receiving the latest livery style three years after other modern classes and it suited them. STD 90 is seen at Golders Green with STD 21 in the distance wearing the old colours. STD 90 was, in fact, the last to receive the red and cream livery and it was also the last to carry it up to the time when it went for its final overhaul in November 1952. The use of offside route number stencils underwent a short term revival and in some cases vehicles from which the holders had been removed now had them reinstated. *Colin Routh*

Having just emerged from overhaul, STD 99 gleams in the latest livery. It is December 1949, the first month in which overhauled vehicles in the Central Bus fleet began appearing from overhaul without front posters already applied. It was quite a novelty at the time to see large areas of sparkling new paint on either side of the destination boxes although, in most cases, these only remained that way for a few days before the bill stickers got to work. *Ron Wellings*

A final overhaul programme commenced in October 1950 from which the whole class eventually emerged in their final livery of all-red apart from a narrow cream central band, a product of London Transport's desire to simplify the painting process to make spraying possible. By January 1953 all one hundred vehicles had been painted into this new, rather unflattering style.

By 1951 the STDs had been the undimmed face of Hendon garage for a full 14 years but, despite the excellent attention that they had received there by the maintenance staff and their continuing good condition, they were now undeniably getting old by bus standards of the time and their replacement by newer rolling stock had to be considered. It had long been London Transport practice for decisions about what types of bus to allocate to each service to be made by a Rolling Stock Allocation Advisory Committee, a joint arrangement between London Transport's management and representatives of the trades unions, and in July 1951 the fate of the STDs came up for discussion. Local Transport & General Workers' Union representatives at Hendon had been pressing for some while for new vehicles on route 13 where they considered themselves disadvantaged by having to operate alongside modern RTs which Cricklewood garage had been using on it since February 1948. Whilst the Union's top brass readily agreed that the condition of the STDs did not justify early scrapping and that they should, in fact, remain in service right up to the final withdrawal of pre-war and wartime buses from the Central Bus fleet, they supported their members' request in regard to route 13 and asked if the STDs currently working on it could be transferred elsewhere to replace G class Guy Arabs. London Transport agreed to this provided that the displaced STDs were kept in a batch at a garage in reasonably close proximity to Hendon. The decision they finally came to was to reallocate the surplus STDs to Enfield to replace Guys on routes 107 and 107A which had a Monday to Friday requirement for 35 vehicles, closely matching the 32 STDs which were expected to become available in the first instance.

Starting with their sixth and last overhaul cycle STDs began emerging from the works in the now-standard 'all red' livery starting with STDs 3 and 7 on 5th October 1950, and before long many of the class could be found carrying this rather uninspiring colour scheme. Typical of this last phase in their working lives are STD 13, STD 39 and one other photographed in the approach road to the old Edgware garage while working on route 113. *Ken Blacker*

Despite their now advanced age, at the start of 1952 the STDs were still too good to scrap. Nevertheless it seemed very strange at first to see some of them redeployed at Enfield garage, well away from their normal Hendon haunts. STDs 31 and 56 were both amongst the batch that commenced work at Enfield on 1st February 1952, all of which were in the 'all red' livery. It soon became clear that, almost to a man, the drivers favoured their speed and handling qualities to those of the much newer Guy Arabs that they replaced. In contrasting scenes on route 107, STD 31 is seen in genteel suburbia while STD 56 – with old-style destination display – passes through trolleybus territory at Barnet.
Peter Mitchell / Alan B Cross

This plan was put into action when, on 1st January 1952, new buses began work from Hendon garage on route 13 and, on the same date 18 of the most recently overhauled STDs were temporarily delicensed. More joined them in the course of the next few days and by the 10th of the month Hendon's full route 13 allocation had been converted. Rather fittingly, in view of the strong Leyland tradition at Hendon, the replacement vehicles were also Leyland Titans in the form of new RTLs, which on Sundays could be found working on all three of Hendon's other services, 113, 142 and 183. However, things tended to happen quickly back then, and after only three months at Hendon the RTLs departed and new RTs were drafted in to replace them.

The transformation of routes 107 and 107A at Enfield took place on 1st February 1952 when 29 STDs were relicensed all at once and were joined by two more transferred directly in from Hendon. Odd ones were received later which subsequently brought the Enfield allocation to a maximum of 36. For the most part, Enfield's STDs stuck rigidly to the 107 and 107A, although occasional appearances in place of the scheduled Guys on route 135 were not unknown and at least one incursion was noted on to the 144B.

By March 1953 London Transport's massive post-war fleet renewal programme had progressed to the point that hardly any double-deckers could be found in London's City and West End other than members of the RT family. Hendon's STDs were still active on route 113, and the only other non-RT types were STLs from Victoria on routes 77 and 77A, plus utility Daimlers and a sprinkling of STLs from Merton, also on the 77/77A plus the 88. A policy was now adopted of replacing the remaining pre-war STDs on a piecemeal basis, a few vehicles at a time, during the course of the next year. As a result, the first casualties occurred at Hendon with the delicensing of STDs 3, 4, 5 and 7 on 1st July 1953, while a start was made at Enfield with STDs 21, 33 and 72 a month later, on 1st August. By the end of 1953 15 of Hendon's STDs had gone, leaving route 113 worked half and half by STDs and RTs but with a full complement of STDs still on the outer suburban 183, while Enfield had lost 20 STDs, leaving just 11 operating into 1954.

March 19th 1954 saw the end of STDs on route 113, by which time just 28 remained licensed at Hendon for the 183, while Enfield's contingent had by then dropped down to a mere five vehicles. The end was now not far off, and Enfield's last two STDs

The final day of STD operation from Enfield garage was 30th April 1954 and STD 65 was one of the last two STDs on that garage's books. Their popularity there had been such that they had been chosen in preference to RTs for working Sunday excursion no 33 to London Airport where STD 65 and one other were photographed at some time in the summer of 1952. They gave a good run for five shillings (2s 6d for children), leaving the garage at 2.30pm and arriving back at 8.27, calling *en route* at Enfield Town, Oakwood Station and North Finchley. *Alan Nightingale collection*

ran on 30th April, being delicensed on the following day. These were STDs 34 and 65. The final demise at Hendon came six weeks later, the last day of operation there being 13th June 1954 when STDs 74, 87, 90 and 97 were available for work on route 183, of which it is believed that STD 90 was the last one back into the garage. After this the only pre-war double deckers still on passenger service in the Central Bus fleet were 14 STLs in the eastern suburbs at Upton Park on route 101, and these ceased operation just over a fortnight later. Their final day at work on 29th June brought the use of pre-war double deckers to an end on the Central Bus fleet. In the Country area the last STLs clung on until 1st September 1954 when the final 24 were withdrawn.

Route 183 had been one of the first Hendon routes to be worked by STDs and it subsequently became the last. Their familiar roar was soon to disappear for ever from places such as Kingsbury Circle which STD 6 negotiates near the end of its working career. Having had the honour of being the very first STD to arrive at Hendon it almost lasted to the very end, being withdrawn just a fortnight before the final day.
Peter Mitchell

STD 83 was one of the lucky ones granted a short reprieve in the driver training fleet which meant that the sight and sound of the Hendon STDs could still be enjoyed, albeit briefly, after their service days had ended. Allocated to various garages (Southall, then Stockwell and lastly Norbiton) during its brief driver training career, STD 83 was one of the very last retained in this role until all were finally delicensed on 25th March 1955.
Roy Marshall

All the STDs withdrawn from service at Hendon were parked up in the large yard in front of Edgware garage which, a few years earlier, had been the collection point for large numbers of brand new RTs and RTLs waiting introduction on tram conversion services. The Enfield fleet were all moved to the traditional storage yard at the old AEC works in Forest Road, Walthamstow. Starting in November 1953 a few were placed into the driver training fleet, the total number being used for this purpose subsequently rising to eleven. However, these, too, were placed on the 'for sale' list and the last of them, STDs 34, 43, 83, 96 and 98, were taken out of use on 25th March 1955.

Even after withdrawal the STDs were still in relatively good condition with plenty of working life left in them, and if things had turned out differently many would have commenced a renewed working career in Yugoslavia. Back in March 1953 London Transport had sold 15 utility Guys and 10 former Green Line 10T10s to the Municipal Authority in Sarajevo, and such was their success that an enquiry was made as to the availability of further withdrawn vehicles. Although London Transport had a large, ongoing contract to sell surplus buses to W North of Leeds, it was established that up to 86 STDs could be excluded from this contract and made available from March 1954 by competitive tender. The Sarajevo authority submitted an offer in late February 1954 for 25 vehicles at £220 each and a month later for a further 60 at the same price, both of which London Transport accepted. Unfortunately, communist authorities at that time were often unable to access sterling funds, even in relatively small amounts, and this happened in the case of the Sarajevo Municipality, causing the deal to fall through.

With the exception of one vehicle (STD 78) which was sold directly to a private owner, the STDs all passed eventually to two dealers, Max Speed Transportation of London SW17 and W North of Leeds who purchased 23 and 62 respectively.

Even in its new life there is no mistaking STD 80, largely thanks to its roof route number box which has now come down in the world! Travelling fairgrounds and circuses were wonderful places to find old buses in the early post-war years, and in the case of STD 80 it passed into the ownership of a showman soon after its initial sale to W North of Leeds in March 1955. The location of the photograph is unknown but the STD is in the company of a variety of showmen's living quarters ancient and modern.

Second hand bus chassis proved attractive to furniture removers because of their low profile and it was by no means uncommon to find them fitted with purpose-built pantechnicon bodies. This newly-rebodied and seemingly unidentifiable example was actually based on the chassis of STD 61 which had ended its working days at Enfield on 1st September 1953 and was subsequently sold to dealer Max Speed Transportation. *J M Aldridge*

What might have been! After a post-London Transport life with Hutfield's Coaches STD 90 ended up in the ownership of Frank Smith of Long Itchington from whom it was bought by the late Norman Day for preservation, and it is seen on the day of its 'rescue'. Unfortunately vandals subsequently damaged it beyond economic repair and it was broken up in 1962. *Ken Blacker*

More than 30 were sold on by them for further use, in a few cases as lorries and pantechnicons. At least 22 found further service as psv's, being particularly favoured for use on works and schools contracts, with notable mini-fleets being six in the service of J Yuille of Larkhall, Lanarkshire, and four apiece with Smith's Luxury Coaches of Reading and Hutfields Coaches Ltd of Warwick and Fareham. Sadly, only one was ever purchased for preservation, in May 1960, but it was scrapped two years later after being very extensively vandalised. This was STD 90, which had initially been one of the Hutfield fleet but was last operated by Frank Smith of Long Itchington, Warwickshire. It is a matter for constant regret that none of these splendid vehicles now exists.

SIX-WHEELED LUXURY – THE LTC CLASS

The LTC class coaches are probably the least well-remembered and least documented of London Transport's pre-war classes. Even in their heyday they managed to keep quite a low profile and the reason for this was that, for the most part, they seldom covered regular duties, so you never knew when you were going to stumble across one. Right to the very end they usually looked well maintained and cared-for, and they were undoubtedly handsome and impressive looking vehicles whose three-axle layout gave an impression – albeit a misleading one – of power and speed. The LTCs were intended specifically as private hire and touring vehicles and, with the exception of the wartime years, this was the role that they normally fulfilled. Their entry into service during 1937/38 marked London Transport's first attempt to create coaches specially designed to serve the top end of the market.

From its formation in July 1933 London Transport maintained a separate Private Hire organisation which came under the aegis of the Country Bus & Coach department but which was administered separately, with its own operating staff and its own fleet of coaches based at its own garage, the one-time Cambrian Landray premises in Waterworks Road, Brixton Hill. The initial private hire fleet was a very mixed one, consisting of vehicles of various makes, shapes and sizes. Few could really be classed as fully modern or particularly luxurious, and many were allocated to the department simply because they had swing doors which made them unsuitable for regular Green Line work. It was not until March 1935, when the task of integrating the numerous private companies into the London Transport system had been completed, that a full internal survey of the private hire fleet was carried out with a view to determining its future size and vehicle composition.

The impressive lines of the LTC are fully illustrated in this view of LTC 22 taken at the Weymann factory prior to delivery. The three-tone green livery adds to its attractiveness, and evidence of careful detailed design is evident in features such as the flamboyant rear mudguards and even the semi-streamlined autovac tank. The double seat visible just behind the door pillar confirms that this was one of the vehicles delivered as 30-seaters from new. *J M Aldridge collection*

At the time it was found to comprise 38 vehicles consisting of no fewer than eight types of vehicle:

AEC Renown	1	LT 1429
AEC Regal	21	T 150-154, 319-324, 393-402
Albion Valkyrie	1	AN 1
Daimler CF6	1	DST 6
Dennis Arrow	4	DL 32-35
Dennis Lancet	4	DT 4-7
Leyland Tiger	4	TR 32-35
Morris Viceroy	2	MS 10, 11

This fleet was adequate in size to cover all off-season requirements but needed to be augmented by a further twelve coaches during the busier summer months. This was usually done by temporarily transferring other vehicles from Green Line work. The continued operation of such a mixed fleet was obviously not efficient or financially desirable, but such were the pressures for fleet renewal on other sections of the organisation that it was almost the end of 1936 before attention became focused on the purchase of a purpose-built and fully modern core private hire fleet, which duly emerged as the LTC class.

When drawing up their specification for its new coaches, the Chiswick design team's goal was to create vehicles that were unequalled for comfort, quietness and smooth running. They also needed to look sumptuous despite lacking the streamlining that was becoming popular with other operators of luxury coaches at the time in favour of a bus-derived dual-purpose style of body not dissimilar to the 50 9T9s built for Green Line work earlier in 1936. As if to prove that they were totally uninfluenced by contemporary trends in the outside world, the Chiswick team produced a quixotic design incorporating both new and outdated features which combined to put the LTC in a class of its own amongst the plethora of luxury coaches vying for trade in London at the time.

One of the quirkiest features of the LTC was that it employed a three-axle chassis in an era when these had largely gone out of fashion (except, of course, for high-capacity six-wheeled trolleybuses). This was done in the belief at Chiswick that the best possible ride quality would be obtained if three axles were used, and there was the further benefit of reduced wheel arch intrusion into the saloon, enabling all seats to be forward facing. Surprisingly, London Transport did not make use of the third axle to specify vehicles of 30ft length to achieve maximum seating capacity; instead it specified a short wheelbase of 16ft 6ins which, in accordance with current Construction & Use regulations, permitted a body only 27ft in length. AEC, who produced the chassis, had to revive its 663 Renown model of short-wheelbase six-wheelers, none of which had been made since the LGOC purchased its last batch of 'Bluebird' LT double deckers back in November 1932. The last of these, LT 1426, had chassis 6631226 and the LTCs followed on as 6631227-1250. The LTC differed from the earlier Renowns in having modern, fully floating rear axles similar to those used on trolleybuses, while the drop-frame extension beyond the rearmost axle was eliminated as it was not needed for coach use.

The LTCs also bucked the trend for modern heavy-duty coaches in being petrol engined. The design team's aim was to achieve a maximum of operating refinement by keeping noise and vibration to the barest minimum, which was only possible using a petrol engine. Economy of operation was not a major factor in this instance, and the probability that a fully laden LTC would be unlikely to achieve much more than 4 miles per gallon was not regarded as important. The LTCs were, indeed, very civilised and smooth vehicles to ride in, and when stationary their tickover was almost inaudible. The petrol engines were not new but had been removed from LT class double deckers, and then thoroughly overhauled at Chiswick before being sent, complete with renovated radiator, to AEC's factory at Southall. As part of the same contractual arrangement, 24 new A165 8.8 litre oil engines were delivered to Chiswick for fitment to the donor LTs, a process which was carried out between May and July 1937 ahead of the construction of the first LTCs. The A145 110mm x 130mm 7.4 litre petrol engines were allied to an axle ratio of 6¾:1, the drive being through the D132 preselector gearbox and fluid flywheel which London Transport employed extensively on STL and other classes.

The contract for building the bodies was given to Weymann's at Addlestone using the same MCW design of metal framework as it had employed on the 9T9s. There were, in fact, many design similarities between the two, although the LTC fortunately lacked the contrived and ungainly built-up bonnet and front wing structure which had spoilt the 9T9s, and it also lacked a rear destination box. Both were embellished by a polished front bumper and both had a split-level waistline to give improved vision for passengers seated towards the rear. The LTCs were not fitted with conventional headlights but instead carried two spotlights mounted lower down on scrolled ironwork. Perversely, to those who did not know why, the LTC's overall length of 25ft 0ins was 5 inches less than the 9T9 even though the latter had only two axles.

It was internally that the Weymann bodywork on the LTCs really excelled. There were only 28 seats and these were wide and high-backed and staggered to provide maximum shoulder space. They had substantial headrolls and armrests and were sumptuously comfortable. On the nearside, two single seats were placed alongside the aperture containing the sliding door, which protruded into the saloon. Behind these were five double seats, while on the offside all eight rows of seats were doubles. Chain-operated wind-down windows, 'Birma-bright' non-corrosive luggage rack rails and castings and a tip-up courier seat mounted on the front bulkhead were all design features of note, but probably the most prominent fitment was the double section 'QT Easiway' opening roof supplied by Quicktho (1928) Ltd of Wandsworth.

The production line in Weymann's Addlestone factory on 16th August 1937 finds an LTC midway through construction with the main body framework in place and cladding work now under way. The number plate 1278 hanging on the radiator is the manufacturer's identification number and indicates that this vehicle is destined to be LTC 5. Another four months were to elapse before the fully-completed LTC 5 was delivered to Chiswick Works on 21st December. *J M Aldridge collection*

Unlike most contemporary coaches fitted with sliding roofs, these opened sideways and consisted of two sections, each 7ft 6ins long with a bridging panel between them. An ultra-modern feature was the fitment of radios (then usually known as 'wireless') to certain of the LTCs to keep passengers entertained or, as the publicity went, "to pick up the world as the coach travels along". The loudspeaker was mounted in the front bulkhead and the radios, which were supplied by Marconi, were housed in the rear luggage rack on the offside and controlled from a master switch in the driver's cab. Although all 24 vehicles were fitted with external aerials mounted towards the rear on each side and along the back of the opening roof structure, Frank Pick was not convinced that radios were a good idea and gave authorisation for only five (although six were actually fitted, on LTC 2, 3, 5, 8, 11 and 17).

The deep seats, staggered to increase shoulder room and fitted with arm rests and ash trays, gave an air of sumptuousness to the LTC. Unlike Green Line coaches, these vehicles were not provided with clocks which London Transport presumably regarded as unnecessary for hire and touring work. On this occasion the roof panels are in the open position allowing light to flood into the vehicle. *London Transport Museum*

The interior of an LTC looking towards the rear and with the roof panels closed. The narrowness of the gangway is apparent but was acceptable in coaches never intended to be used on regular bus work. *London Transport Museum*

Authorisation to order the 24 coaches was given at the Chairman's Meeting on 11th February 1937 at a total cost of £55,000. Registration numbers EGO 505-528 were duly obtained and body numbers 17610-17633 were allocated. Under the Board's classification scheme they were known as 1LTC1. The LTC derivation was, of course, obvious, being a combination of LT (acknowledging that these were basically LT class chassis) and C (for Coach). AEC delivered the chassis to Addlestone between 27th July and 16th September 1937 and the first complete vehicle to be received at Chiswick was LTC 3 on 30th September. The final arrival was LTC 23 on 4th February 1938.

The completed vehicles weighed in at a prodigious unladen total of 8tons 5cwt. Fully laden, they could weigh as much as 10tons 3cwt (compared with 8tons 15cwt 2qrs of a 9T9). They were, in fact, underpowered and as a result they inevitably gained a reputation for being slow and ponderous. With the fuel consumption commensurately high, it is fortunate that they were fitted with 45-gallon petrol tanks. However, painted in the Country Bus department's livery of three shades of green with a single polished moulding below the windows, they undoubtedly looked the part, and they certainly marked a welcome and long-overdue upgrade to London Transport's private hire fleet.

The first seven vehicles delivered were allocated to the private hire garage at Brixton between 18th November and 2nd December 1937. These were LTC 1-3, 6, 8, 11, 12. Their stay there was destined to be short-lived. A major internal re-arrangement meant that control of the Private Hire department was to be transferred from Country Buses & Coaches to the Central Bus department, and the Brixton garage closed with effect from 22nd December 1937. On this date the entire private hire fleet was transferred to the Central Bus garage at Old Kent Road, the takeover by this department becoming fully effective on 24th February 1938. The first LTCs licensed new at Old Kent Road were LTC 5, 13, 15, 18, 19 on 1st January 1938 followed by LTC 20 and 21 on 5th January. These were destined to be the last ones built as 28-seaters.

Almost from the very start it became apparent that a seating capacity of only 28 was insufficient and arrangements were made with Weymann for the final ten (LTC 4, 7, 9, 10, 14, 16, 17, 22-24) to be built as 30-seaters. The increase was achieved by replacing the two single seats at the front nearside by doubles. In order to avoid the already narrow central gangway being narrowed any further, and to fit into the restricted space next to the sliding door enclosure, these seats had to be narrower than the remainder although built and upholstered in the same general style. The ten 30-seaters all entered service during March, between the 24th and 28th of that month.

The 14 earlier vehicles were all converted to the new 30-seat arrangement during March (with some possibly being dealt with in February) to ensure that all 24 LTCs were of the higher capacity in time for the 1938 summer season. Even as 30-seaters their capacity sometimes proved inadequate, so much so that in January 1939 a formal proposal was put forward to convert a 'proportion' of them (the exact number was unstated) into 32-seaters at a cost of £155 per coach. This failed to gain official sanction as it would involve the upheaval of installing a complete set of new seats throughout each coach.

The end of the 1938 summer holiday season saw eleven of the LTCs delicensed on 1st October for the duration of the winter, but all were relicensed for another anticipated busy year on 25th March 1939. For the 1939 summer season the LTCs were joined at Old Kent Road between April and June by new TF class coaches (TF 2-13) specially designed for sightseeing work, thereby completing the renewal of the private hire fleet. Their arrival saw six of the LTCs moving from Old Kent Road to Holloway in May and June, giving a north London location from which to cover some of the touring and hire work. Sadly, fast moving events on the international front meant that this was to be the one and only summer in which the two batches of new coaches would work together. With the outbreak of war imminent all 24 LTCs were taken out of service on 1st September 1939, and within a day or two they had all been converted into ambulances with their seats removed and capacity installed for carrying up to nine stretchers. The Marconi radios were also removed from the vehicles which had carried them.

A typical view of an LTC in service shortly before the outbreak of war brought their activities to a halt. The vehicle in this case is LTC 14 which was one of the last to enter service in March 1938 just in time for that season's Easter trade. The weather must have been hot as all the saloon windows have been fully wound down and the roof panels are fixed in the open position. *D W K Jones*

A stark change of employment from 1st September 1939 found the LTCs fitted out as ambulances. This early wartime view finds LTC 20 with its newly repainted grey roof and regulation white markings, and with half-masking on the windows to shield patients on stretchers from prying eyes. The vehicle appears to have been on active duty when photographed as the windows are open and so are the rearmost roof panels. LTC 20 spent most of the war years allocated to Palmers Green garage. *D Evans © Omnibus Society*

The acceptance of new contracts for touring and hire work ceased immediately and most existing commitments were cancelled except for certain jobs classed as strategic for which a few of the new TFs were retained in service up to the middle of October.

For their new duties as ambulances the LTCs were allocated to a variety of Central Bus garages at which they were serviced periodically and where they stood-by awaiting emergency calls to spring into action. In common with the rest of the fleet, they almost immediately received masked lighting to comply with blackout regulations, and white markings on their mudguards and elsewhere on the bodywork to enhance their visibility after dark. The roofs were initially camouflaged with grey paint although in some cases this was later changed to brown. No LTCs were overhauled during the war and it is not known if any were repainted externally. Certainly most, if not all, were still in their original three shades of green – now mostly very faded – when the war ended. None was lost through enemy action and most led an uneventful wartime career. In May 1944 five of the class (LTC 12-16) were converted back to conventional 30-seaters, presumably for the purpose of carrying patients capable of walking, but retained their status as ambulances. The only break from routine came when LTC 2, 5-7 and 17 were sent on loan on 7th June 1944 (a day after the D-Day landings) to the Canadian Hospital at Wokingham where they were used to convey wounded Canadian troops evacuated from the European War Zone between south coast ports and the hospital for treatment and recreation. A sixth vehicle, LTC 4, remained with London Transport but was stationed at Staines garage (and later Windsor) as a spare. This arrangement ceased on 15th August 1944 when the LTCs all returned to London Transport garages.

The LTCs continued to be in use as ambulances for a short while after the war ended, but on 19th December 1945 all 24 were delicensed and placed in store awaiting a decision as to their future use. During January and February 1946 all those still with stretchers were reunited with their original seats, and a programme of external repainting got under way to smarten their appearance. Earlier, the Board had received a message back in July 1945 from the Regional Transport Commissioner indicating that private hire work could now be got under way without his special consent. There were, however, strict provisos that round trips could not exceed 70 miles (or 50 miles in the case of holiday outings or recreational journeys), and that there should be no interference with the full operation of stage carriage, special workers' or other essential journeys. Initially this relaxation met a negative response from London Transport which was unable to take advantage of it until a marked improvement in the staff position had occurred, which meant that there was no great hurry to get the LTCs back on the road.

In a later view, and with their roofs now painted brown and main headlamp masks removed, ambulances LTC 2 and LTC 5 are parked close to their allocated garage, Chalk Farm, awaiting their next call to work. Almost seven years after it was first applied, their original green paintwork is now beginning to look faded. *R H G Simpson*

In fact, London Transport resumed private hire work in a small way on 1st January 1946 using whatever vehicles happened to be available, no LTCs having returned to service as yet. Demand for bus and coach hire gradually built up, and a limited tour programme was resumed from 11th June 1946 consisting of a morning and afternoon 'Seeing London Tour' and a half-day trip to Windsor. It had initially been planned to commence these on Easter Monday but a dearth of foreign tourists brought about the delay. The pre-war system of housing the coaching fleet at Old Kent Road and Holloway was not perpetuated, and as the LTCs were gradually reinstated into service they were allocated to a variety of garages, probably those with the most favourable staffing position. Although a Private Hire department was re-formed, it was merely an office-based function at 55 Broadway which processed orders and produced itineraries and it no longer had control of any vehicles.

The coach fleet itself was badly depleted, eleven out of the twelve TFs having become war casualties, leaving just TF 9 to soldier on alongside the 24 LTCs. The first LTCs to be put back into public service were LTC 13, 15 at Tottenham garage and LTC 18, 20 at West Green, all on 25th March 1946. April 8th marked a much larger reintroduction encompassing ten vehicles, these being LTC 1-3, 5 at Willesden,

Its six year spell as an ambulance having ended in December 1945, LTC 9 was repainted into green and white bus livery and placed back on the private hire roster based at Old Kent Road garage. In April 1947 it was repainted once more and from then onwards carried Green Line style livery for the remainder of its career. *S A Newman © A B Cross*

With freshly repolished front bumper and newly-applied two-tone green paintwork, Old Kent Road-based LTC 10 is seen on an early post-war private hire job for the Municipal Passenger Transport Association with a second LTC close behind, in this case in the green and white colours. Both still have their radio aerials in place at this time, but not for much longer. *Charles F Klapper © Omnibus Society*

An early post-war revival for the benefit of tourists was the Windsor half-day tour which quickly proved very popular. Victoria-based LTC 14 was photographed covering this duty prior to going for its first overhaul in February 1947. *J F Higham © A B Cross*

The first overhaul for the LTCs, in 1946/47, saw removal of the now-redundant radio aerials, resulting in a completely uncluttered rear end appearance as demonstrated by Camberwell's LTC 6 after it emerged from overhaul in March 1947. It is seen working an Earls Court to Olympia shuttle service in connection with a British Industries Fair exhibition. *Alan B Cross*

LTC 6, 7 at Palmers Green, LTC 9, 23 at Old Kent Road and LTC 11, 19 at Chelverton Road. All had been freshly repainted into the Lincoln green and white colours introduced for Country area vehicles at the start of the war, with LONDON TRANSPORT fleet names. These were augmented by LTC 21 at Chelverton Road and LTC 22 at Palmers Green on 19th April which were the last LTCs to appear in green and white. The remainder, allocated to Croydon, Hounslow and Victoria between 29th May 1946 and 1st February 1947, were all in full Green Line colours but still bearing the LONDON TRANSPORT fleet name.

Despite having been in the fleet for up to nine years, none of the LTCs had ever received a full overhaul and London Transport set about rectifying this omission late in November 1946, with the intake of LTCs for overhaul continuing through to March 1947. Each overhaul was very thorough and took about two months to complete, and during the process their semaphore-type trafficators were removed as were their radio aerials and loudspeakers. Garage stencil and running number holders were fitted. This overhaul programme meant that all the LTCs received a second coat of paint within a matter of a few months, and it resulted in the green and white livery applied to sixteen of them being very short lived, as full Green Line colours were now the standard for the private hire fleet. The last green and white LTC had, in fact, been LTC 15 which went in for overhaul on 19th March 1947.

Transfers of LTCs between garages in the early post-war years frequently happened and were probably usually dictated by the staffing position, space availability and engineering capabilities. It was sometimes possible to find LTCs deputising for double deckers on Central Bus services in the years just after the war when not required for private hire or tour work, but they were not liked by staff. The narrowness of their gangways rendered them unsuitable for constant to-and-fro passenger

It was occasionally possible to find an LTC covering a Central Bus duty at times of dire vehicle shortage. On this occasion Chelverton Road garage has taken the opportunity to put LTC 21 out on route 30 on an occasion when it was not booked for hire work. The passengers on route 30 must have been more than unusually surprised to be supplied with a vehicle with sumptuous coach seats and the front section of its roof wide open. *J M Aldridge*

movement and made the conductor's job difficult, but the greater problem was their inability to run on time. J B Burnell and B H Hawkins, the very influential Operating Managers of the Central and Country Bus departments respectively, were both on record for declaring that LTCs were unsuitable for use on scheduled public services, except possibly on a very few Country Bus routes, because of their low engine power.

The Country Bus department, to its credit, gave the LTCs a fair trial starting with the acceptance on 10th June 1947 of eleven which were allocated to Hertford (4), Dunton Green and Epping (2 each), and Northfleet, Luton and Hitchin (1 each). In most

Hertford garage was the main recipient of LTCs during the 1947-49 period, and though they were sometimes used on its bus roster the main preference appeared to be to keep them on Green Line work. LTC 1 is seen inside Hertford garage ready for duty in place of a 6Q6 on route 715, while on the same day and in the yard at the back LTC 2 is laying over between shifts along with Hertford garage regulars C 44 and T 504. Note that, even as late as 29th March 1948, the Cub still carries the white blackout disc on its back panel. *Alan B Cross*

The need to direct a few LTCs to work as buses or Green Line coaches in the Country area in 1947-49 marked a brief but interesting phase in their career. LTC 10 was available for these duties at Epping garage between August and November 1948 and was photographed covering a scheduled working on route 399.

cases they were used on Green Line work, to which they were better suited than conventional stage operation, although Hertford – and to some extent Epping – also used them on the latter from time to time. All this came to an abrupt halt on 30th June 1948 when they were needed to help out with the Olympic Games contract but ten returned when this was over, in August 1948. This time Windsor garage was added to their itinerary but none returned to Dunton Green or Northfleet. The start of the 1949 summer hiring and tour programme saw their final departure from Country service on 24th April.

The late nineteen-forties pattern of operation for the LTCs required three of them to be available on a daily basis to cover London Transport's scheduled sightseeing tours along with TF 9. The growing popularity of these tours led to unfavourable comparisons being drawn between the LTCs and TF 9. Visibility was regarded as poor from the LTCs, and the deep, high-backed seats were considered unsuitable and, indeed, relatively uncomfortable for this type of work. Perturbed by this, London Transport's Commercial Manager enquired whether three LTCs could be converted to TF 9's standard

The 1948 Olympic Games at Wembley marked a hectic period for the Private Hire department and kept the LTCs busy. Foremost in this view on 4th September are LTCs 14, 12, 11 and 1 while, further long the row, can be found three RTs plus an STL, a CR and a Q The LTCs have all come in from Country garages, LTC 14 from Hertford, LTC 12 from Epping, and LTCs 11 and 1 from Dunton Green. Note that LTC 11 is minus its front bumper; it was quite a rarity to find an LTC in this condition. *J M Aldridge*

of suitability by installing glass cove panels and standard Green Line-type seats. This was duly agreed in January 1949 at a cost not exceeding £500 per vehicle, which represented good value at the time as the LTCs were expected to remain in service until 1954. It was decided that the small scale of the project did not justify detailed drawing office work, and that the conversions should be carried out in the Experimental Department's shop at Chiswick.

LTC 20 was the first of the three to be converted and it was shown off, accompanied by TF 9 for comparison purposes, to the London Transport top brass at 55 Broadway on 2nd February 1949. The two subsequent conversions were LTC 1 and LTC 18. The result certainly marked an internal transformation, with three glass roof panels on both sides providing much additional light and vastly better all-round vision, and Green Line style aluminium-framed seats successfully reducing the air of claustrophobia. The internal luggage racks had been removed and the opening roof panels, though still retained, were now permanently fixed in the closed position.

A much bigger conversion scheme, which this time embraced the whole LTC class, was agreed in June 1949 and put into effect later in the same year. London Transport's rolling stock programme for 1949 envisaged the final withdrawal from service of all petrol engined vehicles in the fleet apart from 163 which needed to be retained (99 LT single deckers, 32 T class, 8 lowbridge ST and the 24 LTCs) and which it was proposed to convert to oil using engines removed from scrapped buses. (This scheme necessarily excluded the eight Inter-station Cubs and the BEA-owned Commer Commandos operated on airport services for which no suitable engines were available for simple conversion.) In the majority of cases it was proposed to use engines from withdrawn STLs, but the LTCs were an exception. In their case the engines were to be A165 8.8 litre units taken from scrapped LTs, which would at last give the LTCs the extra power that they had always needed. The cost of conversion for each LTC was budgeted at £160, which would be quickly recovered by an estimated saving per year of £125 in operating costs.

A May 1949 scene in Chiswick Works looking down at LTC 1 with its newly installed glass roof panels, the light from which reflects on and clearly demonstrates its new Green Line-type tubular framed seats. In the background are three STLs and newly-overhauled G 310. LTC 1 returned to Chiswick later in the same year and stayed there between 12th October and 7th December undergoing overhaul and the replacement of its petrol engine by a diesel unit. At the same time a plain offside front mudguard was installed in place of the original streamlined rubber unit. *J.C. Gillham*

LTC 15 has now taken on quite a pugnacious air with its oil engine installed and its bonnet extended to accommodate the extra engine length. It re-entered service with its replacement engine upon emerging from overhaul on 18th April 1950. Photographed in Camberwell garage yard, its final scheduled allocation, LTC 15 stands in the second photograph next to its nemesis, brand new private hire coach RF 2. It is now Festival of Britain year and two STLs specially retained for Festival special services are in the background. *D W K Jones*

The replacement A165 oil engines were all fully overhauled at Chiswick before being fitted to the LTCs. There was a certain irony in the fact that, twelve years earlier, this had been the type of engine supplied new by AEC for fitting to LTs as a means of providing petrol engines for the new LTCs. Now the reverse was applying and it was the LTCs that were gaining the A165s and, once again, the use of the long 8.8 litre engine meant that the bonnet had to be lengthened by just over 4½ins. Thus, the radiator, also the bumper, on each LTC had to be moved forwards, producing an instantly recognisable snout effect. Inevitably these once almost-silent coaches became considerably noisier and more vibration prone, but now with an altogether more lively performance. The fitment of the oil engines coincided with the second and final overhaul programme for the LTCs. As a result, the first two to emerge as oilers were LTCs 16 and 21 on 4th November 1949 and the final day of operation in petrol-engined form was 13th March 1950 when LTC 13, then based at Putney Bridge garage, performed its last duties.

This final overhaul came at a time when London Transport was busy removing the streamlined and purely decorative rubber rear extension from the front offside mudguard of many vehicles built in the middle and late 1930s. Most of the 9T9s and 11T11s had been treated in this manner as had several hundred STLs, and most of the LTCs were now dealt with likewise. However, one pre-war decorative feature that remained right to the end was the polished moulding below the windows, an embellishment not favoured on any other classes by London Transport.

Back in February 1949 it had become evident, through a major shuffling of new bus orders which relieved Metro Cammell of an obligation to build 550 RT bodies, that this manufacturer would now be in a position to supply bodies for new RF single deckers very much earlier than originally anticipated. The first 25 of these new RFs would be fitted-out for private hire and sightseeing work, exactly the right number to replace the current fleet of such vehicles consisting of the 24 LTCs plus TF 9. It later transpired that the new RFs would become available in time for the main 1951 touring season, although the older vehicles would remain in

Even in their final operational days the LTCs still looked very presentable and it is a great pity that not a single one survives today. Both carrying Tottenham garage plates, LTCs 20 and 22 exhibit several differences from each other. LTC 20 was the first of the trio to be given glass quarter lights and the tops of its low-backed seats are just visible. LTC 22 still has roof panels which open and has retained its old luxury seats, but – in common with most of the class in its latter days – the front offside mudguard has now lost its streamlined extension. *Alan B Cross*

service for a few months longer to cater for the huge upsurge in demand anticipated for the duration of the prestigious Festival of Britain celebrations planned for that year. Even though the early arrival of the RFs meant that the remaining life expectancy of the LTCs was reduced from an anticipated 4 years to about 1¾, the financial outlay of converting them from petrol to oil was still considered worthwhile and, as recorded above, the project was allowed to go ahead.

1951 marked one last, glorious summer for the LTCs which were joined from April, as anticipated, by the new RFs. May and June 1951 also saw the arrival of fifteen new RFW luxury coaches which, though not originally part of the planned fleet renewal programme, were a welcome if un-London-looking addition to the coaching fleet. The end came abruptly on 1st October 1951 (or, in the case of LTCs 10 and 14 at Leyton, 2nd October) when the whole LTC fleet was delicensed, with all but one of them being dumped on spare land alongside Potters Bar garage where they quickly began to look forlorn and run-down.

The story had not quite ended. Customer pressure for hire contracts was such that six LTCs had to be resuscitated for the 1952 summer season, two of them from May and the remainder from 1st June onwards. The vehicles concerned were LTC 1, 9, 10, 18-20 which, perhaps not surprisingly, included the three modernised in 1949 with roof lights and new seats. This, however, really did mark the end of the line for the LTCs and all six were removed from service when the summer season ended on 1st October 1952. LTC 12 had already been sold by then, and the rest were put on the disposal list, passing to dealers between January and October 1953. The only exception was LTC 2 which was retained in stock until it was finally determined that no LTC would be required for London Transport's museum fleet, its final date of departure being 18th June 1954.

Sadly, there was no real afterlife to record for the LTCs. Two or three chassis, without bodies, are thought to have been exported to Spain, but most would have been broken up after surrendering all items of possible further use.

Spare land at the side of Potters Bar garage was where 17 out of the 24 LTCs spent their final days with London Transport, gradually becoming ever more decrepit over a period not far short of two years between withdrawal from service and eventual sale. Apart from one broken window and a stolen radiator badge LTC 15 still looks fairly complete. Most retained their final destination blinds and one of the many enthusiasts who visited the LTCs in their final resting place has probably wound through this one to see what hidden gems it contained. *G A Rixon*

UNDERFLOOR ENGINES – THE TF CLASS

Of all the new types of single decker introduced by London Transport during the pre-war years, the one that undoubtedly deserves the highest accolade for combining the latest and best in design techniques with a justified reputation for speed, smoothness and reliability of performance was the TF. It is such a pity that (with the exception of an early prototype) it was manufactured so close to the outbreak of war that six years were to elapse before it saw proper service and could properly be judged for what it really was. Long before full post-war operations resumed no less than 12.5% of the TF class, which originally totalled 88 vehicles, had been destroyed during the Blitz. Even so, sufficient remained to thrill with their performance in the hands of drivers who clearly enjoyed working on them and knew how to obtain the best from a vehicle which was probably unequalled by anything else built during the pre-war era.

The genesis of the TF can probably be traced back to late 1934, or certainly early 1935, when London Transport's engineers were already in discussion with Leyland Motors over the latter's tentative plans to develop bus chassis designs incorporating, alternatively, a rear or an underfloor engine chassis. These plans happened to coincide with the Board's own strongly held view that the traditional front-mounted engine was not the best arrangement on a single decker. The contractual agreement with AEC meant that the bulk of London Transport's bus and coach purchases would have to be sourced with its long-time supplier, whose Q-type was adopted as the new standard until its shortcomings became evident and an enforced reversion back to front-engined models became inevitable. However, this did not stop the design teams at Chiswick and Leyland from exchanging views and, in due course, from producing prototypes and, later, initiating limited production runs. The TF was the outcome of one such joint design project.

In the body shop at Leyland on 7th May 1937 most of the framework and external panelling for the future TF 1 is complete with details of the driver's cab area being finalised. The fully-glazed half-drop windows have been received from their manufacturer and installed, while a full-sized rear route number box is in position but will soon be removed in favour of a much smaller unit. The position of the engine is in clear view, and in the background stand a couple of fully-completed new STDs. *J M Aldridge collection*

About a week later, on 13th May, further construction has been carried out on the driver's compartment which has now progressed sufficiently to allow this photograph to be taken to demonstrate the unusually high driving position. The floor in front is marked to demonstrate the area of visibility forward of the driving seat. The water filler cap is in place on the front bulkhead but work has yet to start on installing internal wiring or fitting-out the passenger area. In the background STD 89 looks almost complete but it was not delivered to London Transport until 12th June.
J M Aldridge collection

Without any formal contract having apparently been drawn up between the two sides, work on the prototype underfloor-engined coach took place during much of 1936 and the early months of 1937. Leyland's main area of responsibility was to design and build the chassis frame and to provide certain specified running units, and it was also arranged that Leyland would build the body using its own standard methods of construction where applicable but to a design provided by the London Transport team at Chiswick.

The latter's main task was to specify and source braking and transmission units to suit its own specific operating requirements, although it also joined in other aspects of the design process as necessary. Such was its impact that Leyland suggested in July 1937 jointly registering the design of the TF, which was agreed. Earlier, on 22nd February 1937, A A M Durrant had already obtained a provisional patent in respect of the rear axle shaft drive. The prototype chassis became no. 11176 in Leyland's then current numbering series.

The chassis frame, which was designated by Leyland as the FEC (Flat Engined Coach) model, was a simple structure compared with Leyland's Tiger and most other passenger chassis of the day, in consisting of a pair of completely flat and parallel side members intersected at right angles by tubular cross-members and box-shaped body outriggers. Rising slightly from front to rear, it permitted the fitting of a completely flat floor to the passenger saloon allowing all seats to face forward. With the saloon floor unavoidably higher than on a conventional front-engined vehicle, three steps were needed to reach it as against two on conventional vehicles such as London Transport's 9T9. Overall length of the chassis was 27ft 6ins and it had a comparatively long wheelbase of 18ft 6ins with a front overhang of 2ft 11ins and a rear one of 6ft 1in. The front axle was placed in the same forward position as on a conventional front-engined chassis even though it would theoretically have been feasible to set it back by a few feet to accommodate an entrance door at the very front. Especially unusual was the fitting of single rather than double wheels at the rear when there seemed to be no reason for doing so. This is thought to have been at the insistence of London Transport who were currently immersed in this type of arrangement on the AEC Q before its drawbacks related to poor road holding had become fully apparent. Because of it, larger (10.50x20) tyres were needed at the back than the conventional (9.00x20) ones at the front.

The power unit was Leyland's standard, well proven overhead valve direct injection 94bhp 8.6 litre unit with bore and stroke of 4½ins and 5½ins respectively, currently standard on Leyland's Tiger and Titan models and specified by London Transport for its new STD class. The engine was laid on its side, suspended below the chassis frame just behind the driver, with minor modifications to the sump and cylinder heads. Because of the size of the engine, the cylinder heads came close to the offside, with both these and the fuel pump readily accessible from outside the vehicle by lifting a body panel. Air to cool the engine was drawn through a grille at the front of the vehicle where the vertical, centrally-mounted radiator was placed. An extension ran forward from the crankshaft to drive the dynamo, air compressor and heater fan. The engine itself was flexibly mounted on rubber, one mount being at the front of the engine on top of the cylinder block and the other at the rear forming a ring around the cardan shaft. Torque reaction was taken by a chassis member on top of the engine, visible through a trap in the saloon floor, and the engine's movement was limited by two rubber buffers, the whole arrangement giving a very smooth ride.

London Transport's influence became really evident in the design of the vehicle's transmission and braking systems. The gearbox was London Transport's much favoured preselector unit of Wilson design manufactured by AEC (from whom TF 1's gearbox was purchased), operating through a fluid flywheel. The gear selector lever was of the electro-magnetic type with gear selection made by means of a solenoid. Instead of being floor mounted, it was located on the steering column below and to the left of the steering wheel and could, in theory, be operated by the driver's little finger. TF 1 broke further new ground for a London bus in that the brakes were air operated, the foot brake pedal being coupled to an air pressure valve controlling the admission and release of air via pressure cylinders serving all four wheels. Air pressure was generated by a Bosch-designed rotary compressor built by Clayton and driven from the front of the engine.

By the beginning of May 1937 the bulk of the body had been constructed in the Leyland body shop, and from the back of the first bay rearwards it had been fully panelled though not yet fitted out internally. The last part to be constructed, and the most visually stunning aspect of TF 1, was the driver's cab. By any standard it looked like an afterthought and was, simultaneously, both ultra-modern and ill-fitting. With its windscreen and side windows stretching from the waistrail right up to roof level it gave an odd glasshouse-like effect and was quite unlike anything ever seen before.

The designers' remit had been to provide the driver with the best range of vision possible, and their answer had been to put him half in and half out of the main part of the body at such a height that he could see the road surface to within 6 inches of both front wheels, enabling any child or animal immediately in front of him to be observed immediately. To achieve this, the cab projected partly into the saloon from which the driver entered it through a doorway containing a sliding door, there being no door for him on the offside. A steep step took him up into the cab followed by a shallower step up to his seat. With his eye-line being above cantrail level, it was necessary to insert a secondary window into the bulkhead behind him above normal window level so that he could peer back into the saloon if necessary. The steering wheel was positioned about 18 inches higher than normal, and the unusually tall steering column and its accompanying linkages all had to be specially made. Because of the driver's abnormally high line of vision, the customary canopy to the left of the driver, which in the normal way would have accommodated the Board's standard destination screen, had to be omitted. Instead, a much shallower and far less visible one-line box was incorporated into the set-back nearside bulkhead. At London Transport's insistence a box of similarly small proportions was also installed at the rear of the vehicle. This required the removal, before it had ever been used, of a normal-sized screen with which the main bodywork had originally been built, giving a sure indication that design changes were being made right up to the last moment.

The completed vehicle poses for Leyland's official photographer on 13th July 1937 with more than a month still to go before it was handed over to London Transport on 21st August. The paintwork and signwriting is now complete although the unladen weight has yet to be ascertained. Although it does not carry a fleet number it is equipped with garage stencils for the Brixton private hire base and fictional running number 100. *J M Aldridge collection*

Excluding the driver's cab, the main bodywork on TF 1 contained all the usual Chiswick influences also apparent in the 9T9 and LTC classes, including windows radiused at all corners and half drops wound up and down by top-mounted handles, plus a raised waistrail at the rear, added to which were a few extra features intended to give an air of greater modernity, such as the downward-sloping window line at each end of the main saloon and flamboyant mudguards. Internal décor was in the usual shades of green and all 34 seats were forward facing, including a pair ahead of the saloon doorway which enjoyed a superb view of the road ahead. The normal sliding door was provided and, as usual, an emergency door was placed in the centre at the rear of the vehicle. A design feature unnoticed by the travelling public was that the body floor was mounted directly on to outriggers from the chassis instead of on the customary underframe, resulting in a form of integral construction which meant that the body could not easily be removed during overhaul as was London Transport's usual practice.

TF 1 had been fully built by 13th July 1937 which was the date when Leyland arranged for official photographs of it to be taken. Its exact delivery date to Chiswick is not known, but acceptance tests were held there on 10th August when it probably looked magnificent painted in its three-tone Green Line livery. The vehicle weighed 6tons 11cwt 3qrs, and with Leyland's recommended rear axle ratio of 5⅖:1 it was found to be fairly sprightly and capable of reaching 30 mph within 32 seconds when accelerating through all gears and carrying a load equivalent to 34 passengers plus 5 standing. It had already been officially decided that, subject to satisfactory performance, the vehicle would be purchased from Leyland for a maximum of £2,500 and payment of this sum was duly authorised. TF 1 was officially taken into London Transport stock on 21st August 1937 and was licensed as DYL 904 on 1st September. It was allotted body number 18075 and given the classification 1TF1.

The other-worldly appearance of TF 1's cab area, which hinted at having been designed as an after-thought, must have really surprised passengers at the time and even today looks somewhat bizarre. This early service view shows the vehicle taking on passengers at Victoria on route from Woking to Tunbridge Wells. *D W K Jones*

In contrast to its startlingly different exterior, the inside of TF 1 held few surprises and followed very closely in terms of general fittings and décor the standard Green Line pattern already well established on other classes. However the driver's compartment was of interest and the accompanying photo shows its sliding door, the two steps by which access was gained to the high-mounted seat, and also the gear selector console mounted on the left side of the steering column.

Prior to its acceptance into stock, A A M Durrant had already announced that some aspects of TF 1's design did not conform to the Board's standard but that these would be rectified in any future orders that might be placed. A particular case in point was the vehicle's turning circle which, at 59ft, was well outside the 56ft considered as the acceptable maximum for a two-axle vehicle of its length. Durrant proposed, and it was agreed, that a prolonged period of testing should be carried out before TF 1 was put into service.

The Chiswick-based test programme went on for about three months and by the start of November confidence in the robustness of TF 1 had grown to the extent that it was officially shown off to the trade press, who had been aware of its existence for some while but had so far put off writing about it. By 24th November 1937 the time had arrived for it to be placed into service and it was despatched to Tunbridge Wells garage for type training prior to entering service on Green Line routes C1 (Tunbridge Wells–Chertsey) and C2 (Tunbridge Wells–Woking). Service operation commenced on 1st December and TF 1 must have made a huge contrast to the elderly petrol-engined T-types which it operated alongside. These were replaced on 1st May 1938 by brand new 10T10s, and the comparison between these and the TF must have been interesting to observe.

Inevitably, with a vehicle as unusual and advanced in design as TF 1 things went wrong once it was exposed to the rigours of daily operation which would not necessarily have been noticed during testing at Chiswick. The air braking system proved particularly troublesome, especially in regard to the Clayton-Bosch reciprocating rotary compressor, and minor weaknesses in items such as the gearbox support brackets and the air cylinder fixing had to be put right. The response by drivers to the ultra-high driving position has gone unrecorded but subsequent events tend to indicate that it was not altogether popular. The lack of a clearly identifiable front destination screen was quickly seen as a major drawback and within a few months of entering service (the actual date is unrecorded) the front of the bodywork was rebuilt at Chiswick to insert a full-width canopy with a conventional destination box incorporated into it. Because of the need to retain the original high windscreen, the front of the box was forced to project well above the main roof line with a pronounced downward slope behind it. At the same time the nearside front bulkhead window and the nearside saloon window ahead of the door were both increased in height up to cantrail level to give the driver improved vision when looking backwards from his high perch to observe events adjacent to the doorway, and the exterior of the bodywork was repainted into the now-standard Lincoln green.

TF 1 had been in service at Tunbridge Wells for only a few weeks before it was officially proposed, on 10th January 1938, that the purchase of 88 TFs should be included in the 1938/39 new vehicle

The exact date of TF1's major frontal rebuild is not known, but it was certainly within only a few months of its initial entry into service at Tunbridge Wells. Now resplendent in the latest Lincoln green, it was photographed outside the experimental shop at Chiswick which is almost certainly where the body reconstruction would have taken place.

The Chiswick drawing office produced this illustration of how the three-axle TF that was planned but never built might have looked. The waistrail, with its drooping window line at either end, is clearly influenced by TF 1, and in order to site the doorway approximately in the centre of the vehicle it has been arranged to make it forward sliding. The length of the rear overhang is notably much greater than on the two-axle TF.

A 'what might have been' if Brush Coachworks had been successful in gaining the order for twelve coach bodies. It is interesting to see how Brush intended to incorporate London Transport's specification for deep windows, upright pillars and roof lights into its own vision of how a modern, attractive sightseeing coach should look. *London Transport Museum*

Leyland's advertisement announcing the purchase of 87 more "flat-engined coaches" as published in 'Passenger Transport Journal' on 8th April 1938. TF 1 was depicted in its original form, but it may well have been at about this time that it was back at Chiswick having its frontal profile drastically altered.

programme and that a tender should be issued to Leyland for the provision of the chassis for these. The figure of 88 was arrived at by including 75 coaches to complete the rolling stock renewal programme for the Green Line fleet, after which no further vehicles should be required for it until about 1945, 12 coaches to special sight-seeing specification for the exclusive use of the Private Hire department and one 30ft-long Green Line vehicle with two front axles capable of seating 38 which would only be built if permission could be obtained from the Metropolitan Police for it to operate on all services experimentally. By 14th February a tender price of £1,150 per chassis had been received from Leyland, and on 10th March the purchase of 88 coach bodies was approved. (In fact, the final price per chassis turned out to be £1,124 which was actually lower than the tendered price, although well in excess of the £944 paid to AEC for the 6Q6s).

Things moved quickly from here onwards. First, the notion of producing a six-wheeler was abandoned (presumably the Metropolitan Police had not proved as accommodating as hoped), so the new chassis requirement was reduced to 87, which led Leyland to make a public announcement on 8th April 1938 that London Transport had placed an order for 87 flat-engined coaches. A construction contract for the 75 34-seat Green Line bodies was awarded to the Board's own factory at Chiswick, and the decision was taken to seek tenders for the 12 33-seat sightseeing bodies from outside contractors specialising in this type of work.

London Transport issued a general outline of what was required for its sightseeing coaches and five well-established coachbuilders quoted for them. A summary produced on 11th November indicated that three out of the five (Brush, Duple and Harrington) had quoted for composite framed bodies, and only Weymann and Park Royal had proposed metal construction which London Transport considered to be desirable. The lower of these two quotations at £958 per vehicle was from Park Royal and this was accepted. The total cost of the 12 bodies would be £14,100 but this included provision for one special vehicle at extra cost equipped with a wireless set and removable seats. This figure was considerably in excess – by £3,300 – of the amount initially allowed for these vehicles in the appropriate Special Expenditure Requisition but as savings were expected on the 75 bodies being produced at Chiswick it was decided that placing of the order should nevertheless go ahead. It is interesting to note that, in finality, the Park Royal bodies actually cost £1,135 each, vastly in excess of the tendered price, but this is thought to have been largely as a result of subsequent modifications required by London Transport. In comparison, the 75 Chiswick-built Green Line bodies eventually cost £897 each.

From April 1938 onwards various designs and models were produced at Chiswick for the 75 forthcoming Green Line coach bodies with special emphasis on establishing a definitive front-end design and a satisfactory layout for the entry steps. The final design was approved on 8th August.

Meanwhile, at Leyland some important chassis design details were put in hand to improve on TF 1 and to overcome some of its shortcomings. A priority was to secure a reduction in the swept turning circle, and with this in mind the wheelbase was shortened by one inch to 18ft 5ins, with a corresponding increase of ½ inch in both overhangs, front and rear, while the front track width was reduced from 6ft 6ins to 6ft 3½ins. The concept of an ultra-high driving position was abandoned as was the single wheel arrangement at the rear. To meet London Transport's latest requirements the use of compressed air to operate the brakes was now extended to also include the preselector gearbox with an air-pressure cylinder mounted inside the gearbox in place of the usual spring, the aim being to achieve much lighter pedal operation. This latest technological advance had already been tried out briefly in service between July and October 1938 on ST 1140, the forerunner of the RT class, but this was the first instance of full air operation being adopted on a production batch and it was not without its risks. A further modification eliminated the column-mounted selector console used on TF 1 in favour of a normal floor mounted lever, while a minor change was made to the back axle ratio to improve operating performance, bringing this to 5⅕:1.

In the Chiswick body shop a pre-production

The front end design adopted for production TFs was an improved and more attractive, albeit slightly more bulbous, version of that on TF 1, and incorporated an attractive stylised bullseye motif on the bonnet top. The first Green Line coach to be built was originally numbered TF 2 although this was later changed to TF 14, and this view of it taken adjacent to the skid patch at Chiswick Works illustrates the neat lines of the Chiswick-built bodywork. *Alan B Cross*

specimen for the Green Line batch was produced which received body no.1. London Transport had decided to commence a new series of body numbers for all future new vehicles starting with the TFs, of which the Green Line batch would carry bodies 1-75 followed by the sightseeing coaches as 76-87. On 28th November 1938 the skeleton for body no. 1 was despatched to Leyland's works to enable the design and position of outriggers, access hatches and other chassis features to be finalised. While at Leyland, it was mounted on the first chassis to be produced, 301625 (the entire batch being 301625-301711). The completed structure arrived back at Chiswick on 19th December for finishing off and received fleet number TF 2 although, as explained later, this was subsequently amended to TF 14.

Following the successful acceptance of body no. 1, a stock of completed Green Line bodies rapidly built up at Chiswick, coming off the production line well ahead of the receipt of new chassis which were delayed because Leyland, like all major vehicle manufacturers, had been required to divert a large part of their attention to the war effort which was rapidly gaining impetus with an outbreak of hostilities looking ever more likely. New chassis started to arrive at Chiswick, roughly in number order, on 31st December 1938, and though only two more were received in January 1939, a steady inflow was achieved from February onwards. So as not to delay construction of the sightseeing coaches, the first two chassis to be received from Leyland were despatched to Park Royal on 9th and 12th January. In anticipation of their arrival, newly-built and as yet unmounted Green Line body no. 34 was sent from Chiswick on 6th January for the Park Royal staff to study and measure, particularly in reference to chassis mounting arrangements, and it was not returned to Chiswick until 21st June after construction of the twelve new sightseeing bodies had been completed.

The Chiswick-built bodies were mounted completely haphazardly as far as their body numbers were concerned, no doubt in accordance with whereabouts in the factory they had been stored and how easy it was to gain access to each one when required. Unusually for the normally well-organised licensing department at Chiswick, confusion appears to have reigned over the fleet numbers to be applied to the TFs. The next three vehicles to be completed, all with Chiswick-built Green Line bodies, were allocated fleet numbers TF 4 to TF 6 (it is not known what TF 3 was intended to be: perhaps one of the Park Royal vehicles?). Eventually it was established that the sightseeing coaches would be TF 2 to TF 13 and the Green Line batch TF 14 to TF 88, and on 13th February 1939 the existing TF 2, 4-6 were renumbered TF 14-17 respectively. None had yet entered service by this date but the original TF 2 (later TF 14) had been licensed as a private car for training purposes from 1st February onwards.

When constructing the bodies for TF 14-88, the Chiswick coach factory employed the same unusual method of construction as it had used successfully when building the 266-strong 10T10 class, combining pillars and crossbars, plus floor

In terms of fitments and general décor the Green Line TFs followed closely the large and popular 10T10 class. The front end layout was, of course, specific to the TF and this view shows the popular front nearside seat and also the driver's door arrangement. This photograph was not taken until May 1947 by which time the clock had been replaced by a bullseye motif. The printed notice on the glass behind the driver's seat refers to the availability of cheap day returns and weekly tickets. *London Transport Museum*

TF 37, which entered service on 1st May 1939, is fully fitted out for operation from Romford garage on route Y, but appears in fact to be on what is probably a press demonstration run. The rear of one of the new sightseeing TFs is just visible in front of it. *Charles F Klapper © Omnibus Society*

and roof bearers, of timber flitched with steel inserts, with waist and cantrail sections made of metal, the whole structure being held together by brackets rather than a series of interlocking joints. In fact, the end result had much in common with the 10T10 although, with a completely flat floor in the passenger saloon, the need for a stepped waistrail was eliminated. The front end was where the TF took on the distinguished and modernistic appearance for which it was renowned, the final combination of slopes and curves having been arrived at after the various tentative plans and models had been examined and commented upon over a period of months in 1938. To the nearside of the driver's cab the radiator filler cap was located centrally with a liftable, sloping panel adorned by a stylised bullseye motif below which was located the cooling/heating system and also the horn. Under the driver's windscreen a removable panel gave access to various items of pipework behind which were the steering column and box. Internal décor and fittings were much the same as on the 10T10 and, although the nearside front bulkhead panel was sited further back than on prototype TF 1, there was still room behind it for the popular seat to be located ahead of the entrance. All 34 seats were in forward facing pairs and, as on the 10T10s, the emergency door was centrally situated in the back wall of the vehicle.

TF 14-88 were given the classification 2TF2 and were officially recorded as having an unladen weight of 6tons 11cwt 3qrs. The accuracy of this, however, is highly suspect, and when a typical vehicle – TF 42 – was selected to be the subject of extensive testing at Chiswick between 1st and 6th May, a big weight discrepancy was discovered. The vehicle, and by implication all 75 2TF2s, was found to be heavier than expected and unable to carry a full load of 34 seated plus 5 standing passengers without infringing the legal laden weight limit of 9 tons. London Transport was sufficiently worried about the problem to despatch TF 14 to Leyland's works in Lancashire on 19th May where the manufacturer took the unusual step of separating body and chassis to seek the source of the problem. TF 14's body (body no. 1) was subsequently re-used in June on TF 75 while, after its return to Chiswick, TF 14's chassis received new body no. 50 on 1st September. This was the only example of a TF chassis being separated from its 'unit construction' body. In early post-war years the declared unladen weight of all the 2TF2s was adjusted upwards to 6tons 15cwt which is probably what it had been all along.

Park Royal began delivering the sightseeing coaches, coded 2TF3, on 30th March 1939 starting with TF 2. The remainder all arrived during April and May with the exception of 'odd man out' TF

The sightseeing TFs were amongst the most modern-looking coaches on the road Their purely functional lines were enhanced by just the right amount of cosmetic streamlining, with a pronounced downward swoop to their pale green paintwork at the front and just a hint of a slope at the back. At the time it was new, in April 1939, no-one would have dreamt that TF 9 would be the only one of the twelve to operate in service for longer than one single summer season. *London Transport Museum*

12 which completed the order on 9th June. The latter was the wireless-equipped vehicle capable of being converted into a 25-seater plus tables for use, when required, as the directors' saloon. These were striking-looking vehicles enhanced by the way that the light green relief swept down at the front to emphasise their modern appearance, with just the hint of a downward slope at the rear. The two shades of green were separated by polished mouldings. Very deep side windows reaching up to cantrail level were surmounted by roof corner lights on each side. The only destination box was a single-line one at the front and, unlike other TFs, the emergency exit was located on the offside behind the rear wheel. As in the earlier case of the LTCs, the whole batch was equipped with wireless aerials which on the TFs were mounted a little untidily on the front dome, presumably in the expectation that one day these might be needed on all the vehicles and not just on TF 12.

Internally the 2TF3 was almost the exact opposite of the supremely comfortable but slightly claustrophobic LTC and it had the advantage that, with 33 seats, its carrying capacity was higher. With its large expanse of glass, and low-backed Green Line-type seats, everything about it was superbly designed for sightseeing and on fine days the roof centre could be slid open for even greater visibility, with the component pieces stacked neatly at the back. The bright airiness inside the saloon was enhanced by the pale green of the rexine-covered window surrounds and all seats faced forwards in pairs with the exception of a row of five across the back. The entrance to the driver's cab was enclosed by a hinged door instead of the sliding variety on all other TFs, and though no luggage racks were included, pockets on the seat backs were provided to carry small items. At a declared unladen weight of 6tons 13cwt, the sightseeing TFs slightly exceeded the declared unladen weight of the Green Line ones, but this was of no significance as their lower seating capacity and the absence of any likelihood of carrying standing passengers resulted in a laden weight well below the legal maximum.

The sightseeing coaches were the only ones in the TF class not to have their emergency door at the back; instead it occupies the last bay on the offside. As with the LTCs, there was no indicator box at the rear. TF 5 is seen, with its roof wide open, on a private hire job during the summer of 1939. *Charles F Klapper © Omnibus Society*

With all 33 seats facing forward and none of the seat tops rising above the window line, an impressive sense of openness was achieved. The moquette seat covering is believed to have been of a design unique to these vehicles, and even the back corners of the body are padded in a complementary though not identical style. Note the light coloured window surrounds and the trendy lamp shades mounted on each pillar. Even when the roof was closed, as on the rearward view inside TF 7, the tall side windows and multiple glazed cant panels ensure that ample daylight streamed into the interior. The view inside one of the batch when in service, and with a tour guide giving his commentary, shows the marvellous all-round vision enjoyed by all passengers when the roof was in the open position. *London Transport Museum*

The arrival of TF 2-13 completed the modernisation of the private hire fleet which now consisted solely of TFs and LTCs, in marked contrast to the chaotic and slightly run-down appearance that it had presented half a decade earlier. The twelve were registered FJJ 603-614 and all entered service from Old Kent Road garage between 2nd April and 10th May 1939, well ahead of the summer high tourist season, with the exception of TF 12 which followed on 16th June. Their introduction into service was not entirely trouble free and between 22nd June and 28th August four of them (TF 2, 3, 11 plus TF 7 twice) had to be returned to Park Royal under warranty for roof repairs, which probably indicates that they leaked.

Most of the sightseeing coaches did not last long enough for many photographs to be taken of them during the few brief months that they were in service. Indeed some may not have been photographed at all. In Westminster, the driver of the lowest numbered of the batch, TF 2, appears to be checking the water level while taking a break during a private hire job. *Alan Nightingale collection*

The Green Line coaches were registered FJJ 615-674 (TF 14-63), FJJ 761-777 (TF 64-80), FXT 41-48 (TF 81-88), and all were destined for Romford garage and its very busy trunk services Y1 (Brentwood–Aldgate) and Y2 (Upminster–Aldgate). Earlier plans to convert these two services to double deck operation using 50 new STLs had now been permanently abandoned, and the TFs were expected to perform better than the 10T10s currently operating them, which would be transferred to other garages to replace older Ts and thereby complete the updating of the Green Line fleet. As a forerunner to the arrival of the new vehicles, TF 1 was reallocated from Tunbridge Wells to Romford on 16th December 1938 and remained there while the new vehicles entered service starting with TF 15-26 on 25th March 1939. No further examples were received at Romford during April, but from 2nd May right through to 5th August new arrivals were steadily recorded and by the latter date 73 out of the 75 were in service there.

The new vehicles were well received by passengers and staff by virtue of their smooth running, quietness and speed, though all did not go entirely well at first especially in regard to the air-operated braking system. A systemic problem also came to light that was to blight the TFs throughout their time with London Transport. The air pressure cylinder incorporated inside the gearbox in place of the usual main spring was arranged in such a way that it was used to disengage gears rather than to engage them, which was the system employed on the RT. This meant that if a vehicle was inadvertently left in gear overnight and, at the same time, the air pressure leaked off, it was impossible to disengage the gear to select neutral when starting the vehicle up. It led to a number of garage incidents when vehicles jumped forwards (or backwards according to the gear currently engaged), with obvious danger implications.

It was mentioned above that two TFs had not entered service by the outbreak of war. These were TF 76, which for some unknown reason remained unlicensed until 16th September 1939, and TF 14 which, though it finally arrived at Romford on the very day that all Green Line services were temporarily withdrawn, 1st September, was transferred out almost immediately afterwards to start its wartime career as an ambulance.

The prospect of converting 300 Green Line coaches to ambulances at 24 hours' notice had been floated with London Transport by the Home Office as early as July 1938, and by 19th September of that year the necessary equipment had been designed and the feasibility of carrying out the conversions successfully tested on vehicles at several garages. The 2TF2s were, of course, not even in existence when this scheme was envisaged but were subsequently included within the ranks of vehicles scheduled to be converted should this need ever arise. With war now imminent, all 75 were duly prepared for their new role as ambulances on 1st and 2nd September 1939, initially with two seats and provision for carrying up to ten stretchers. Subsequently, in February and March 1942, the necessity for providing better arrangements for the walking wounded saw 20 of them (TF 15, 21, 22, 24, 26-33, 36, 37, 42, 45, 59, 61-63) modified to hold 18 seats with space for only four stretchers.

The vehicles were now under the command of the Home Office but remained under London Transport control for day-to-day servicing and for the provision of crews. They were allocated to a variety of garages, and inter-garage transfers of ambulances occurred fairly frequently throughout the war years to suit London Transport's maintenance and staff arrangements. To quote one typical example, TF 59, the garages to which it was attached during the wartime years were: 2/9/39 Battersea, 10/9/39 Old Kent Road, 29/2/40 Seven Kings, 17/2/42 Tottenham, 7/2/44 Palmers Green.

Having no emergency door at the rear, TF 2-13 were unable to accommodate stretchers and this made them unsuitable for conversion to ambulances. With the imposition of blackout regulations the glass in their roof lights was painted over and as outstanding private hire contracts were fulfilled, or cancelled as being unnecessary for the war effort, the twelve coaches were gradually transferred away from Old Kent Road to Country area garages, principally Addlestone and Guildford, for temporary use as buses. Six (TF 2-4, 6-8) left immediately for the Country area on 1st September and the last, TF 12, on 13th October. By 25th November

A line-up of six TF ambulances awaiting their call to duty early in the war with TF 18 in the lead. These are just a few of the fifteen TFs allocated to Chelverton Road garage at the time which, due to shortage of space within the garage, had to be parked on various local streets. It is not known what the DANGER sign on the front of TF 18 refers to, but it may simply have been nothing more sinister than a deterrent to stop inquisitive people or potential vandals from boarding. *D W K Jones*

they were no longer required for bus work and a meeting was convened to determine their future. It was decided that such hire contracts as remained could be covered by other vehicles and that TF 2-13 should be stored for the time being and their future decided later. For convenience, they were all collected back into Old Kent Road garage between 24th and 29th November 1939, where the whole batch was delicensed on 1st December. On 22nd and 23rd February 1940 they were all removed for long term under-cover storage, along with many other delicensed vehicles, in the old Thomas Tilling works at Bull Yard, Peckham (the site, in post-war years, of the splendid new Peckham bus garage). As is well known, tragedy struck in the form of a German bomb on the evening of 7th September 1940 rendering most vehicles stored there completely irreparable and resulting in them being officially written off books on 22nd October. Of the twelve private hire TFs, only TF 9 was salvaged. The remaining eleven, all with less than six months' service to their credit, passed into history almost unnoticed, having had little opportunity to even start making an impact on London's coaching scene.

The only members of the class to see normal passenger service during the war years were TF 1 and TF 9. The same proud driver was photographed with both of them, probably during the period in 1942 when Merton was working the Heinemann contract at Kingswood with TFs before it went over to double deckers. Grey paint covers the glass roof panels on TF 9. *Derek Fisk collection*

By sheer chance the prototype TF 1 happened to be out of action at the time war broke out. On 14th August 1939 it had been scheduled to go to Chiswick Works for its first overhaul, which turned out to be a prolonged affair lasting very nearly ten months. While it was there, the opportunity was taken to reconfigure the front end to bring it in line with the production batch. This required fitting a new, more bulbous nearside front-end structure and new mudguards similar, though not identical, to all other TFs. At the same time, for a reason that is not now known, the diesel fuel filler cap was repositioned closer to the rear nearside wheel which may indicate a relocation of the fuel tank itself. A floor-mounted gear selector lever similar to those on standard TFs replaced the column-mounted unit and the steering wheel was lowered by an inch or two, but otherwise no general reconfiguration of the driver's cab took place. These modifications had all been carried out by 3rd June 1940 on which date TF 1 was recorded as "standardized" and recoded 2TF2. This recoding was, to say the least, arbitrary and appears to have been made without a full understanding of what the coding system was meant for, since the chassis still retained its original one-off specification of single rear wheels on the back axle while the body continued to differ from those on TF 14-88 in very many ways.

After being overhauled TF 1, which now carried standard green and white bus livery, was not immediately required for service and remained delicensed, eventually finding itself, along with TF 2-13, in store at Bull Yard from 12th June onwards. On the night of the fatal bombing, 7th September, it was one of the few vehicles – along with TF 9 – which were saved. Having been one of the last to arrive for storage at Bull Yard, it was probably amongst the easiest to extricate, after which it was taken to Old Kent Road garage for storage. A few days later, on 11th September, it was relicensed there to allow it to carry out special authorised private hire work. This comprised a regular commitment, then requiring two vehicles, to serve William Heinemann's huge Windmill Press printing works at Kingswood on a contract which had been inherited by London Transport in November 1933 when it compulsorily acquired the Golden Arrow business of E Puttergill Ltd, and which the Private Hire department had operated ever since. On 5th October 1940 TF 1 was joined on the Heinemann run by TF 9, which had now been restored to operational fitness, and these were the only two TFs to see passenger service during the first half of the nineteen-forties.

On 23rd February 1942 the two operational TFs were transferred from Old Kent Road to Merton, which may have marked the date when the northern terminus of the Heinemann contract was switched from central London to Morden Underground station. A month later, on 25th March, the contract was revised to be operated by a double decker and the two TFs were reallocated on this date to separate destinations. TF 9 was sent to Hornchurch, where it provided a daily connection from east London to the British National Film Company's Elstree Studios at Borehamwood in connection with vital war work carried out there by the BBC.

TF 1's new operating base was Sidcup garage but details of the contract on which it was employed are not known. It was subjected to a second overhaul, this time at Chiswick tram depot, in November and December 1942, but difficulties in obtaining spare parts for it under wartime conditions resulted in it being permanently laid-up on 1st April 1943. TF 1 then spent the rest of the war unlicensed in Old Kent Road garage and, as soon as authority to dispose of redundant vehicles was given by the government, it was put up for sale.

TF 1 left London Transport on 21st January 1946 upon its sale to Chelsea-based dealer Henry Lane and swiftly found a new home in the Castle Coaches fleet of C C Grundon based in Lewisham. Like another well-known Leyland prototype, the former C 1, it subsequently became well known for its regular appearance on London Transport bus services when the hiring of privately-owned coaches commenced in October 1947 to provide peak-hour assistance to the Board's own hard-pressed fleet. Operating from Catford garage, it could usually be found on route 1 but sometimes on other services until its hiring ended in August 1949. Castle Coaches kept TF 1 going until its end came in July 1954 by which time it had reached the ripe old age of 17, a remarkable achievement for a vehicle that had started life as an experimental one-off.

After the hires to London Transport of 1947-49 ceased, Castle Coaches continued to find plenty of work for TF 1 and even added a LEYLAND badge to the front panel above the registration plate. An enthusiasts' visit by the Southern Counties Touring Society to the Hastings Tramways Company in July 1953, the month after London Transport ran its last TFs, found the vehicle in good condition albeit now 16 years old, and still capable of a good turn of speed. *John C Gillham © Tony Peters*

Meanwhile TF 9 soldiered on in service at Hornchurch, receiving its first overhaul in October 1942 at which time the pale green areas of the bodywork were repainted white. In March 1945 its duties at Hornchurch ceased and on the 25th it was removed to Chiswick ready to be spruced up for its return to sightseeing work in the post-war era which now clearly lay ahead. TF 9 was finally relicensed at Old Kent Road in December 1945, and on 11th June 1946 it moved to Victoria garage, which was destined to be its permanent post-war home, for a resumption of the sightseeing tour programme. At this stage in its career it still carried its never-used wireless aerial.

By June 1946, when this photograph of TF 9 was taken, it was regularly employed on the Seeing London Tour. Being the only one of its kind still in existence it was one of the most instantly recognisable vehicles in the whole London Transport fleet and it could frequently be spotted carrying tourists around the streets of the Capital. The paint had now been removed from its roof windows but it still carried a few wartime markings such as the masked side lights and the white-painted door post, and it also retained its radio aerial, but these were all removed when it went for overhaul in February 1947. *S L Poole © A B Cross*

The TFs came back into regular Green Line use with the introduction of the 723 on 6th March 1946. Like all of the original Grays garage contingent, TF 73 was newly repainted into green and white bus livery before returning to service. It is seen carrying these colours, and also the original black on white Green Line destination blinds, at the Aldgate terminus where its crew are making use of the mobile staff canteen nearby. *J F Higham © A B Cross*

The 75 Green Line TFs all survived the war and, starting on 12th December 1945, they were gradually stood down from their ambulance duties to the extent that, by 1st May 1946, all except eight were either in store or in process of being prepared at various locations for their return to Green Line work. None had been overhauled during the war and, with insufficient facilities available to commence an overhaul programme immediately, the best that could be done was to give them an external repaint. As with the LTCs, the first few were painted into green and white bus livery but from April 1946 the standard post-war Green Line colours were used. Up to ten are known to have been green and white (TF 18, 60, 64, 65, 70-73, 76, 79) but there may have been a few more. They were no longer needed at their original base in Romford as pre-war plans to double deck its services had now been revived, albeit not with STLs but with brand new semi-utility D class Duple-bodied Daimlers. Alternative garages and suitable services had to be found for them.

St Albans-based TF 87 displays standard Green Line colours as it rounds Marble Arch early in the post-war era. The real objective for the photographer was probably the vehicle accompanying it, LT 1165, which Cricklewood garage has borrowed to avoid a 'No Bus Available' situation on route 16 in the absence of a suitable double decker. The missing front wheel disc and side route board, and general lack of cleanliness on TF 87 are signs that the inside staff at St Albans were under pressure and put the vehicle into service in a hurry.

The first TFs to resume Green Line service after the war did so at Grays garage on route 723 (Tilbury–Aldgate). Ten vehicles were licensed there at first, on 1st March 1946 for a 6th March service commencement, but this figure had been increased to 20 by the end of the year. The next TF-worked routes were 714 (Dorking–Baker Street via Kingston) and 727 (Luton–Kings Cross via Barnet), which both started on 1st May 1946 for which Dorking and Luton garages both received nine TFs apiece. Finally, on 29th May 712 (Dorking–Luton) and 713 (Dorking–Dunstable) were introduced, bringing a further allocation of TFs to Dorking and introducing them to St Albans garage for the first time. These two services differed in routeing from the 714 and 727 in that they travelled via Epsom and Morden south of the Thames, and via Hendon and Elstree north of it.

The TFs settled into their new surroundings very well, although complaints about them abounded during their first winter when it was found that the heating system, both in the driver's cab and in the main saloon, failed whenever the water level in the radiator was allowed to fall, even by a small amount. This was a drawback not encountered during their previous spell in Green Line service, back in 1939, which was terminated before the cold weather arrived. It was a problem peculiar to the TFs and while it did not affect the cooling of the engine it meant that drivers had to be sure to replenish water levels regularly throughout the day.

Right from the start it became commonplace to find the Green Line TFs also operating on local bus services from all four garages. These workings were mostly unplanned, last-minute substitutions for other vehicles although, in the case of Grays garage, there were scheduled TF workings on routes 371, 374 and 375. Their use as buses became even more pronounced as the eight remaining ambulances were gradually released and returned to service, in each case after a full overhaul. The final two to serve as ambulances, TF 24 at Hackney and TF 27 at Chelverton Road, were delicensed from these duties on 5th July 1948.

At senior management level the use of TFs on bus services was considered to be a wasteful use of resources, and on 12th January 1949 the situation was largely rectified when an additional Green Line service was added to the TF roster. This was route 720 (Bishops Stortford–Aldgate) which involved introducing TFs to a fifth garage, Epping. Eleven were gathered at Epping, with all of the other four TF garages making a contribution and, coincidentally, providing Epping with a consecutive run of eleven fleet numbers, TF 30-40. Epping garage also covered Green Line route 718 (Epping–Windsor) which it shared with Windsor and, inevitably, TFs stated to appear occasionally on this too. They also appeared regularly on certain positioning journeys on bus route 396. The choice of the 720 was an inspired one. With their greater turn of speed compared with the 10T10s formerly used on it, it was found possible from the start of the summer programme on 1st May 1949 to speed up the running time and to remove a complete vehicle from the 720 schedule to considerable financial advantage.

In the garage yard at Grays TF 26 has covered a scheduled service bus run on the 371 prior to picking up its main duty on Green Line 723. Repainted from its early post-war livery of green and white into full Green Line colours in March 1948, it still looks shiny six months later in contrast to Bluebird-type 'learner' bus ST 1076 standing beside it. *Alan B Cross*

With none of the Green Line TFs having been overhauled during the war, an almost lethargic and certainly spasmodic programme for making good this omission was introduced from March 1946 onwards, by which time the vehicles were all six years old or more. It is not known how or why each individual one was selected to go in for overhaul, but 21 were taken into the works in 1946, 25 in 1947, 20 in 1948 and 9 up to June 1949 by which time all 75 had been dealt with. From July 1949 a series of second overhauls commenced which finally embraced 39 of the class and came to an abrupt halt when TF 31 was sent in for overhaul on 26th June 1950, subsequently becoming the last to emerge on 19th October. In between times, body repaints were often carried out, the dates of which remain unrecorded, and this ensured that the TFs always looked in presentable condition and a credit to the Green Line title.

TF 9 was also overhauled during the post-war years, on two occasions. The first was between February and May 1947, during which its unused radio aerial was removed. The second overhaul, between 21st March and 1st June 1949, resulted in a massive change of appearance to the vehicle through the adoption of a completely new livery style. It was now painted in two-tone green, in Green Line fashion, with the lighter shade confined to the window frames, except at the rear. The distinctive down-swept swoop was eliminated and the lower of the polished aluminium mouldings was overpainted in green. The end result was to give TF 9 a much more mundane and considerably less striking appearance than it had borne in the past.

About a month after the conversion of route 720 to TF operation, Epping-based TF 31 appears to be carrying almost a full load as it heads for its northern terminus at Bishops Stortford. The two male passengers in the front seat enjoy the splendid forward view for which these vehicles were so well renowned.
S L Poole © A B Cross

TF 9's final overhaul, from which it resumed its role on the Seeing London Tour on 1st June 1949, found it wearing a livery style similar to that of the LTCs and the Green Line fleet. It could still be easily recognised by virtue of its unique shape but with the loss of its streamlining it now stood out from the rest much less well that it had done before. On this occasion the open roof sections have been stacked up near the front instead of neatly at the back as was usually the case.
J H Aston

With its end nigh, the Southern Counties Touring Society hired TF 9 for one last sentimental trip before its inevitable departure from the fleet. Brand new RF 19, then based at Middle Row garage, joined it to provide a contrast between old and new. *Geoffrey Ashwell*

With the arrival of the nineteen-fifties, plans for introducing the new RF class were getting under way and, as recorded in an earlier chapter, the first 25 were earmarked to renew the existing private hire and touring coach fleet consisting of the 24 LTCs and also TF 9. All 25 new RFs were licensed between April and June 1951 in good time for that year's main tourist season. Furthermore, the previously unplanned 15-strong RFW class had appeared at exactly the same time as RF 1-25 and theoretically the LTCs and TF 9 were now all redundant. It was thanks to the massively popular Festival of Britain in 1951 that all the pre-war coaches survived the onslaught of new rolling stock and continued to remain in use until the end of the 1951 tourist season, when TF 9 was delicensed and placed in store at Reigate. A surge in demand for sightseeing work brought a brief revival of its fortunes in 1952 when it returned to service at Victoria throughout the month of May, but the surrender of its licence on 1st June 1952 finally marked the end of TF 9's London career. Stored in Finchley trolleybus depot for almost eighteen months awaiting its fate, it was eventually sold to W North of Leeds on 2nd December 1953. TF 9 had outlasted its eleven compatriots in London Transport ownership by a full 13 years. North's are rumoured to have sold TF 9 to an overseas buyer in 1954 but no details of this transaction or its subsequent history (if any) have come to light.

1951 marked the last full year of TF operation on the Green Line network. In September 1951 the definitive list of conversion dates to RF operation was drawn up and, with new deliveries ultimately running almost as scheduled, this programme was adhered to without any need for further amendment. The first four sets of conversions were all focused on routes operated by 10T10s or 6Q6s and it was not until January 1952 that TFs were expected to be involved with the allocation of new RFs to Grays garage. Meanwhile, the immediate future for the TFs had become a matter of some dispute between the engineering and operating management on Country Buses. The former wanted to make their immediate withdrawal a top priority because of mounting difficulties over the procurement of spare parts, whereas the operators wished to retain them in service until all the 30-seater 10T10s had departed to take full advantage of their higher seating capacity. The latter prevailed, and it was decided that the bulk of the TFs would be transferred to bus work while a few would be dismantled for cannibalisation to get over the spare parts problem.

TF 35 was one of the last of its class on scheduled Green Line work, but when route 720 at Epping was converted to RF on 1st May 1952 it was demoted to bus work at Hitchin. It is seen a month later at the busy St Mary's Square bus stand on a 'fill-in' short working to Stevenage on route 303. Other vehicles glimpsed on the stand at the time include an RT, a Cub, and a Birch Brothers Leyland PD1. *Ken Blacker*

Meanwhile, two events occurred in 1951 to impact upon the operation of the TFs just before the end of their Green Line career. On 4th July 1951 a new route 723A was introduced between Tilbury and Aldgate as an offshoot of the Grays-worked 723 to serve the massively expanding residential developments at Belhus Park Estate and North Stifford. Then, on 30th September 1951, route 714 was extended northwards from its central London terminus at Baker Street via Kings Cross to cover the routeing from there to Luton of the 727 which was withdrawn. Luton garage now gained an allocation on the 714 and its crews began covering new territory (for them) southwards to Dorking.

At the start of 1952 the deployment of the 75 Green Line TFs was: Dorking 22, Grays 21, Epping 10, Luton 10, St Albans 10, Romford 1 with one unlicensed. (Romford garage was allocated the odd single decker from time to time for staff, private hire, and service relief work.) However, change was imminent and on 4th January the first TFs departed from Grays for a new, albeit brief, life at St Albans on local bus work. The clear-out at Grays continued throughout that month and its last TF left on 1st February.

The remaining conversion programme to RFs proceeded from the dates shown as follows:

1st March – Dorking, Luton (714)
25th March – Dorking, St Albans (712, 713)
1st May – Epping (720)

Scheduled TF bus workings at Grays were henceforth covered by post-war 15T13s and at Epping by 10T10s. No precise date is known for the ending of TF operation on scheduled Green Line work but the withdrawal of TFs 39 and 40 at Epping on 1st June marked the last possibility of this happening. However, TFs could still spasmodically be found running on the Green Line as reliefs.

For their last spell of scheduled service, which was expected to last only until the middle of 1953 when new Country Bus RFs would become available to replace them, the TFs were concentrated at a group of garages in the North Division of Country Buses consisting of Hertford, Hitchin, Luton, St Albans and Hatfield. Between them these garages covered a large number of rural services, in some cases interworked between garages and with many examples of vehicles switching from one route to another during the course of their day's work. Luton and St Albans were, of course, already TF operators while Hertford and Hitchin received their first allocations of the class on 1st March 1952. Hatfield had a requirement for only one single decker to work scheduled duplicates (building up to an allocation of two from 1st September 1952), and it received a TF for this on 5th May 1952.

Having been downgraded to bus status, the TFs gradually had their GREEN LINE fleet names replaced by LONDON TRANSPORT ones and, very surprisingly in view of the short life-span ahead of them, at least 20 (i.e. almost one quarter of the existing stock) were given complete exterior body repaints into the now-standard 'all green' bus livery. A proposal to replace the popular double seat ahead of the doorway with a single, inward-facing one to provide greater space for the conductor to stand in and to allow the fare chart to be repositioned from the emergency door to the front of the vehicle was turned down on cost grounds, but it was decided that the saloon heaters should be disconnected (although this may not actually have taken place on all of them). The need to break some vehicles up for spare parts resulted in TF 73 being the first of its type to be withdrawn, on 8th April 1952, whilst TFs 22, 32, 39, 40 were also dismantled at Chiswick between May and September of that year with any unwanted remains being sold for scrap.

The number of TFs still in stock exceeded the then current requirements and at 1st January 1953 ten vehicles were in storage unlicensed. The 60 still operational on this date were allocated as follows: St Albans 28, Hertford 16, Hitchin 7, Luton 7 and Hatfield 2. This was acknowledged to be only a temporary arrangement, the prediction being that new Country Bus RFs would begin entering service as early as March 1953. The first of these would go to Reigate to displace the last remaining outpost of 4Q4s, after which the TFs would be the next to succumb.

The final days of the class are not clear-cut, the picture being confused by the festivities in June for the coronation of Queen Elizabeth II. Coronation Day itself was fixed as a public holiday for Tuesday 2nd June but celebrations, including decorated buildings and festive lights, were expected to continue throughout much of the month right up to the 21st. Anticipating the possible need for special sightseeing services and extensive duplication on a number of Green Line routes, it was decided to keep some TFs licensed right through to the end of the month even though they would be redundant by then from their regular bus duties. In line with the replacement programme, St Albans began losing its TFs on 17th April 1953 and Hertford on 11th May, and while most departed into storage, odd ones were retained for possible Green Line duplication. Hitchin's TFs joined the exodus on 1st June and Luton's on the 5th, but it was probably the withdrawal of Hatfield's pair of regular TFs on 22nd June that marked the real end of their short spell on bus operation, although even there a coronation extra remained until the end of the month.

It is not known why a vigorous repainting policy was pursued on the TFs in 1952 with such a short life span projected for them, and their reappearance in the drab 'all green' livery did nothing to enhance their looks. A typical recipient of the new colours was St Albans-based TF 70 photographed in 1953 passing through Borehamwood. *Buses of Yesteryear*

The contrast in latter day livery styles is demonstrated by TF 72 (now in 'all green') and TF 74 (in traditional colours) on a 1952 excursion from Brentwood to an unrecorded destination operated from Romford garage. Romford had no single deckers of its own at the time and the two TFs had been borrowed from St Albans. Standing beside them is T 701 from Grays, a vehicle which eventually outlasted the TFs in the London Transport fleet by a few months. *D A Jones*

On the morning of Coronation Day, 2nd June 1953, TF 14 prepares to leave Amersham garage to take a trip on route 336 in place of the usual RLH. It is not known how many of the TFs licensed for the Coronation month were actually used in service, and in the case of TF 14 this may well have been the first and only occasion on which a TF was ever operated by Amersham garage. *Ken Blacker*

Meanwhile, in expectation of increased demand because of the coronation a TF had been reintroduced at Dorking and another at Grays from 1st June whilst Epping received no fewer than four, presumably with the possibility in mind of them being required for huge augmentation of route 718 which served Windsor. Most surprisingly, two garages that had never previously operated TFs received one apiece, namely Amersham from 20th May and Addlestone from 1st June. It is not known how much use these coronation extras were put to and it is possible that some were used only very spasmodically, if at all. As the accompanying photograph shows, Amersham used TF 14 on Coronation Day itself, while Epping garage is known to have run TF 58 on route 718 on 8th June (the last specifically recorded example of a TF on Green Line work) with TFs 64 and 66 on local bus service on 12th June. By 22nd June a few were definitely no longer required and were sent for storage at Garston, still licensed but probably unused, and when all remaining TF licences were surrendered on 1st July 1953 their service on London Transport was at an end. For the record, the last ones licensed were: Addlestone TF 19, Amersham TF 14, Dorking TF 46, Epping TF 58, 64, 66, Grays TF 83, Hatfield TF 20, Hitchin TF 47, St Albans TF 72. Licensed at Garston right up to the end, but almost certainly not needed, were TF 17, 29, 48, 49, 53, 69, 84, 88. The TFs had simply drifted out of use, with no record ever kept of when, or where, or on what day, the last one actually ran in passenger service. Most unusually for a class of such successful and highly regarded vehicles, their passing went virtually unnoticed.

No time was lost in finding buyers for the TFs. The first eleven withdrawals, after short-term

After demotion to bus duties the odd TF occasionally reappeared on Green Line work from time to time in substitution for a missing RF. On this occasion Luton garage has put TF 53 out on route 714. This vehicle had, in fact, been at Luton for Green Line work since May 1946 and remained there on bus duties when displaced from the 714. Unfortunately, having been newly painted in the latest drab livery, it fails to fully recapture the glamour of former Green Line days. *Allen T Smith © A B Cross*

storage in the operational garages at Hertford, St Albans and Romford, were moved temporarily to the former Kingsway tram subway until they were sold in May 1953. The remaining 59 were all gathered together in the almost-new Garston garage, which had opened for operation on 18th June 1952 but was as yet not fully used. Most of these were moved out in August and September 1953, the purchaser of all except one of the 70 vehicles that remained to the end being W North of Leeds. The exception, TF 77 (which had last run at Hitchin until being finally delicensed on 1st June 1953) was retained at Garston for the museum fleet. It was formally transferred to this on 9th April 1954 and moved to the Reigate garage annex to join London Transport's other museum vehicles.

North's managed to sell a few TFs for conversion into lorries but the demand for such unconventional chassis was not great. None found renewed work on psv duties within the UK, but a couple of non-psv oddments were TF 24 which worked as a tar sprayer and TF 52 as a mobile shop. The only known exports (although there may well have been others that have gone unnoticed) were one (at least) which operated, still with original bodywork unmodified, in Ceuta, Spanish Morocco, and three which found their way to Malta. The first of these, whose LT identity remains unknown, entered service on the Valetta-Rabat route in 1954 carrying a new, locally built body but still in more or less original mechanical condition with its front wheel discs in place as a giveaway feature to its original identity. Two others

By arrangement with dealer W North of Leeds London Transport arranged for certain single deckers to be modified for right hand running by moving the entrance door to the (British) offside. TF 83 and an unidentified 4Q4 are parked in Chiswick Works painted in purplish-pink and yellow for future service in Spanish Morocco in July 1953 prior to conclusion of their sale to North's in the following month. *John C Gillham © Tony Peters*

Recognisable as a TF only by its front wheel discs and longer-than-normal wheelbase, this unidentified TF began running on the Valetta to Rabat service on the island of Malta in April 1954. The new, locally-built body was in the then-current blue and cream livery of the Rabat buses. This vehicle was later rebuilt with a front-mounted engine and ended its days adorned with a Bedford SB grille. *Peter Trevaskis*

entered service in 1957/58 (one in Malta itself and one on the neighbouring island of Gozo) but these, in typical Malta fashion, had been heavily rebuilt and now had front-mounted engines as well as new bodies, and the original one on the island was also subsequently converted to this format. The last of the three to be withdrawn, as late as 1999, was the former TF 83. It had the honour of being the final TF to run in service anywhere, albeit with probably very little apart from its chassis frame remaining to connect it to the vehicle that first entered service at Romford sixty years earlier.

Preserved TF 77 remains today part of the London Transport Museum reserve collection and is normally out of sight in its premises at Acton, although occasional trips out serve to remind us of what splendid and attractive vehicles the TFs were back in their heyday.

The only TF to survive up to the present day is TF 77 which is normally housed in the London Transport Museum's store at Acton. It is seen dressed with replicas of the blinds and boards that it carried for a few months in 1939, during a visit to the Brooklands Museum in October 2018 on the occasion of a Leyland-themed 'TransFest' event organised by the London Bus Museum. *John Waters*

ENGINES AT THE BACK – THE CR CLASS

Following hot on the heels of TF 1, another remarkable vehicle arrived at the gates of Chiswick Works on 16th October 1937 in chassis form. The second of a pair of avant-garde designs to emerge during the year resulting from the joint labours of the skilled design teams at Leyland Motors and London Transport, CR 1 was the test-bed for a planned intake of new-type 20-seaters needed to complete the modernisation of the Board's fleet of small one-man operated buses. Within London Transport, the existing C class vehicles were already referred to colloquially as 'Cubs' in deference to the type name bestowed upon them by Leyland, so it was only natural that the new 20-seater would be referred to by Chiswick as 'Rear Engined Cub', hence the CR classification (Cub Rear). The same title was used at Leyland, who appropriately classified it as the REC model, and it was destined never to be built for anybody other than London Transport.

At the time that CR 1 was being conceived and constructed the conventional front-engined Cub was still very much in production (and continued to be until the outbreak of war), so it might be wondered why London Transport chose to opt for an untried and much more expensive design with which to complete the modernisation of this section of its fleet when, by extending the C class, it could have achieved the standardisation it always desired. The answer seems to be that A A M Durrant and his team were obsessed with pushing the boundaries of design to their maximum, and to justify building the CR the Board's publicity on the matter asserted, somewhat tongue in cheek, that it was built "to improve conditions on one-man Country services and to overcome certain disabilities normally associated with buses in which the driver collects the fares. The normal bonneted type of vehicle suffers from certain drawbacks in internal accommodation and in inferior visibility for the driver".

Carrying Leavesden Road (WT) garage plates and St Albans route stencils for route 355, CR 1 attracts attention from passers-by as it waits outside the St Ermin's Hotel, close to 55 Broadway, on 10th January 1938 for inspection by London Transport's top brass four days before officially entering service. Accompanying it for comparison purposes is C 6, a Northfleet-based vehicle. *D W K Jones*

The genesis of CR 1 can be traced back to January 1936 when Durrant reported to the Board that Leyland had produced an experimental rear-engined single decker and that negotiations were being entered into with the Company to acquire one experimentally. In fact, Leyland's rear-engined bus was purely a development project which was not pursued to its completion: its bodywork was never fully fitted out and it never entered service. However, Leyland was clearly happy to take the concept of developing a rear-engined bus one stage further, albeit in co-operation with London Transport, with the latter's engineers taking the lead in the project. The CR was the ultimate result of this collaboration and in its finished form it differed considerably from the Leyland prototype. The latter had been based on a full-sized chassis capable of carrying around 40 passengers, and had a set-back front axle and single tyres all round, while the large 8.6 litre oil engine was mounted transversely at the rear with the gearbox located in the offside back corner. The CR, on the other hand, was a small 20-seater with normally positioned wheels, including doubles at the rear, and with the engine and gearbox mounted in line with the transmission.

Progress on the CR was such that by 23rd March 1937 Durrant was ready to submit a memorandum seeking official approval to purchase the experimental chassis at a cost of £550, plus a further £650 for a Chiswick-built composite 20-seat body to be mounted on it.

Leyland allocated chassis number 17407 to the new rear-engined Cub chassis. Employing a wheelbase of 14ft 9ins (9 inches shorter than that of a standard Cub), it was powered by an engine mounted longitudinally on the centre line of the rear overhang, and forming a unit with the worm drive casing and a special constant mesh gearbox. The hollow wormshaft of the final drive was sandwiched between the engine and the gearbox. The arrangement was such that the whole unit could be withdrawn from the vehicle through the back or, alternatively, the clutch and gearbox could both be removed separately. The engine, which was rubber mounted, was the same 6-cylinder H2 direct injection model as used on the Country area Cubs with the same power output. Instead of a conventional back axle, the De Dion suspension system was used, employing a universal joint at both the wheel hubs and the differential and using a solid tubular beam to hold the opposite wheels in parallel. This system, which was particularly popular in racing car technology at the time, was claimed to keep the wheels vertical and thus more in contact with the ground as the suspension rose and fell. The prominent rear hubs of the De Dion type axle were a striking feature of the CR and, in designing the drive system, patents held by Streamline Cars Ltd of Maidenhead were used with their approval. This company, though no longer a car manufacturer, existed to exploit patents covering design work done by its principal shareholder, Sir Charles Denniston Burney, a gifted designer and pioneer of motor car, airship and torpedo development,

A modernistic but rather fussy-looking body, number 18001, was constructed at Chiswick in the closing months of 1937. It had a more conventional front end than TF 1 but, like that vehicle, it failed to make use of the vacant space next to the driver except to provide a storage compartment for the batteries. This meant that the passenger entrance, which was enclosed by a G D Peters electro-pneumatic door, had to be situated behind the front wheel. As a result, passengers boarding the vehicle were nowhere near the driver; in fact, they were separated from him by an inward facing seat for two placed immediately ahead of the doorway on the nearside and a transverse seat for two on the offside. The driver's access point to the cab, through which he was also supposed to take fares, was just in front of the first offside transverse seat, no doorway having been provided for him in the offside of the bodywork. From the point of view of the driver/passenger interface, the whole front-end layout was clearly a retrograde arrangement compared with the proximity of the driver to the entrance door on the standard bonneted Cub and a very obvious design weakness in a bus specifically intended for one-man operation.

A further 16 seats were provided to the right of the entrance door, all facing forwards, with the final three rows being placed at a raised level. At the

CR 1's official photographs were not taken until after it had worked at St Albans for a short while and the traffic notice visible through the front windows – which is actually pasted on the glass behind the driver's seat – announces forthcoming entertainments in the town in February and March 1938. A separate polished bumper was fitted at the back of CR 1 but there was never a matching one at the front. *London Transport Museum*

very back was a substantial shelf over the engine on which it was intended that luggage could be stored. Obviously, the emergency door could not be located at the back and instead it was placed about half-way along the offside. As CR 1 was intended for operation by the Country Buses the exterior was painted accordingly in the then-standard livery of three shades of green with black lining-out.

At 22ft 7½ins in overall length, CR 1 was more than 1ft shorter than the C class, but its height of 9ft 0½ins exceeded it by 2 inches and made it look bigger and more substantial. An unladen weight of 4tons 10cwt marked an increase of more than half a ton, so a revised back axle ratio of 5½:1 (instead of 6½:1) was adopted to maintain performance. CR 1 became a complete unit when the new body was mounted on to its chassis sometime during November or December 1937, and in cold and wet conditions on 28th and 29th December it undertook a series of acceleration and braking tests at Chiswick both unladen and laden to the equivalent of a full load of 20 passengers plus 5 standing. Next day, on 30th December 1937, the vehicle was officially taken into London Transport stock and duly registered as ELP 294. Its debut in front of London Transport's directors and other dignitaries took place at 55 Broadway on 10th January 1938.

The front end of CR 1 was more aesthetically pleasing in appearance than that on TF 1, and with the driver seated at a conventional height a full-sized destination box and conventional canopy were both possible. The lower front panel has been removed for the photograph and the opened flap on the bonnet top reveals that the compartment below it contains nothing more than the batteries. *London Transport Museum*

CR 1's working life commenced on 14th January when it made its appearance at St Albans garage. The reason for choosing St Albans is not known and it seems to have been a strange choice because, within a few months, this garage was scheduled to lose the bulk of its one-man operations leaving just a handful of 20-seaters to cover a much-reduced rota of rural services and schools journeys. Documentation no longer exists to record CR 1's performance during this early stage of its career but it is remembered as having evoked the ire of drivers, who often had to walk through the saloon to take fares because passengers had failed to realise that they were meant to pay on boarding. Another cause of driver dissatisfaction arose because it proved harder to achieve a satisfactory standard of gear changing than on the Cubs and Bedfords due to difficulty in synchronising changes on the crash box when the engine was at the back of the bus and difficult to hear. Gear changing when ascending hills was a particular problem especially if pulling away from a stop was involved. By all accounts CR 1 was off the road as often as it was on, and it is known for a fact that it was sent to Leyland's main factory in Lancashire on 26th May 1938 for investigative and remedial work, returning back to London just over a month later on 30th June. After just over a year at St Albans, CR 1 was delicensed on 23rd January 1939 and taken back to Chiswick for a further set of tests to be carried out.

The poster carried by CR 1 advertising "Out and About", a Green Line coach guide, seems a little inappropriate on a vehicle which, by all accounts, spent much of its time broken down. Standing over a pit in St Albans garage, with its lower front panel removed and the engine compartment doors wide open, CR 1 was certainly not ready to go out and about on the occasion that this photograph was taken.

A view through the tall rear doors reveals little more than the radiator and filler cap. The engine is mounted in front of it, in line with the chassis, and the large corner panels are arranged to swing open to gain access to it. *London Transport Museum*

Meanwhile, back at head office things were moving apace to secure the production of a fleet of CRs. Without waiting to see how CR 1 performed in service, A A M Durrant proposed in a memorandum to the Board dated 6th January 1938 that 73 CRs should be purchased during the financial year 1938/39 with the proviso that this figure might be reduced by 15 if certain proposed service revisions were implemented. He was given the go-ahead to seek a quotation from Leyland. Frank Pick took a personal interest and decreed a number of changes to the body design from that of CR 1, including revision to the side windows so that they should all be on one level and the placement of the front registration number plate in a central position instead of towards the offside below the windscreen. By 21st March, Durrant was able to report a quoted price from Leyland for 60 (later reduced to 59) chassis at £525 each and, with Chiswick-built bodies costing £660 each, the total price per vehicle would be £1,285. Permission to place the order was granted with completion stipulated by June 1939. As a follow-up, it was agreed on 6th April 1938 that a royalty of 1½% on the manufacturer's chassis price would be paid to Streamline Cars for use of its patents. In finality each vehicle cost a total of £1,339, an increase over the original estimate of 4.2%, mainly caused by the bodies proving considerably dearer to construct than originally anticipated.

Scenes inside CR 1 show the inward-facing seat for two ahead of the doorway and the raised section for twelve at the rear. The positioning of the fare board on the safety screen adjacent to the emergency exit and the deep shelf at the back above the engine are also evident, but the most notable feature is the remoteness of the driver's compartment from the entrance door which is a sure recipe for inconvenience in the taking of fares. *London Transport Museum*

A new operating base was found for CR 1 at Windsor garage where it worked on route 442 from March 1939 until July 1940. It is seen outside Slough station shortly before the outbreak of war. *J F Higham © A B Cross*

In July 1938 London Transport issued a public announcement that said, "Following exhaustive tests at St Albans garage of an experimental vehicle, London Transport have placed an order for 59 rear-engined Leyland Cubs" and went on to describe CR 1 in some detail. This was widely reported in the lively transport press of the day, none of which picked up on the fact that the order had been authorised before CR 1 had turned a wheel in service and not as a result of 'exhaustive tests' as the press release had implied. Many within the industry would have been aware that CR 1 was not performing at all well at St Albans, but it was not expedient for trade journals dependent upon manufacturers' advertising and operators' goodwill to disclose unpalatable facts such as this. Not long afterwards it was realised that even the reduced number of 59 new CRs would prove excessive in view of the constant drive towards replacing 20-seaters with larger vehicles and Leyland was asked to reduce the order to 48. On 6th September, Durrant was able to report that Leyland had not only agreed to the reduction but had also offered to keep the price per chassis the same despite the reduced quantity, for which he was asked to convey the Board's grateful thanks.

On 25th March 1939 it was time to put CR 1 back into service, but this time at Windsor garage where its performance continued to be closely monitored. Because of the difficulties with fare collection, the driving staff at Windsor would allow it to operate only on route 442 on which it worked the same specific duty each day. A log was kept which recorded wide discrepancies in the number of days per week that it was available for service, with miles run per week over a period of four months varying between 87 and 869, and miles per gallon ranging wildly from 9.3 to 14.4 which must have been quite disappointing, as standard Cubs regularly averaged about 16.7mpg. The need for brake modifications saw it proceed to Leyland's at Kingston on 18th April from which it returned to Chiswick on the 21st, resuming work at Windsor on 1st May. Early engine failure occurred on about 20th July after about 9,000 miles of operation with two seized pistons, badly scored main bearings and a broken fan. And so it went on. Between 20th November and 30th December 1939 CR 1 was away at Chiswick for its first and only overhaul, from which it returned in the new Lincoln green and white livery, and on 30th July 1940 it was delicensed, remaining in store at Windsor for the remainder of the war.

With the general air of uncertainty affecting all forms of manufacturing in the months leading up to the outbreak of war, it was not surprising that the deadline of June 1939 for the new CRs to be built was missed. However, Leyland worked on the assembly of the chassis at Kingston during the late spring and early summer, resulting in the delivery to Chiswick of the first one, CR 2, on 19th July 1939. A further thirteen (CR 3-15) had been received before war broke out on 3rd September. In terms of general design the chassis were very similar to the prototype, incorporating from new various modifications that had been made to CR 1 as a result of its service trials. A notable difference was that the engines supplied for the production batch were of the later, slightly more powerful but definitely less reliable 3½ins x 5ins indirect injection type already encountered on C 77-98. Leyland chassis numbers for CR 2-49 were 201969-202016 and they were allocated registration numbers FXT 108-155. London Transport's classification for the 48 vehicles was 2CR2, following on from CR 1 which was the sole 1CR1.

Although there was little mechanical difference between CR 1 and the production batch, big changes to the bodywork had been made as a result of service experience. Only the back elevation of the body remained virtually unchanged. The front was subtly different, due mainly to a revision of the curved structure to the left of the driver's cab and a lowering of the cantrail moulding, but the sides were entirely re-designed. The biggest drawback of CR 1 had been its front-end internal layout, which rendered it almost entirely unsuited to the one-man operation for which it was intended. For CR 2 onwards the nearside front half-bulkhead was moved rearwards and placed level with the back of the front wheel, while the passenger door was moved forwards to meet up with the bulkhead. By this means the seats ahead of the doorway were eliminated and passengers boarding the vehicle were now closer to the driver, although the drawback still existed that he had to turn almost completely around to take money and issue tickets which, though a big improvement on CR 1, was still a much less satisfactory arrangement than on the traditional Cubs.

The relocation of the entry door meant that all the body side pillars were now positioned differently from those on CR 1 while, in accordance with Frank Pick's wish, all the side windows were now on one level. An interesting feature was that by moving the position of the door forwards it impinged on the decorative flair at the rear of the front mudguard, which was now attached to the door and opened with it.

CR 9 was taken into stock nine days before the war began and joined others of the production batch at Kingston, where it entered service on 10th October. Before doing so it was photographed carrying all its wartime regalia and fully fitted out for service on route 216. The notice on the window behind the doorway reads "Please pay as you enter and tender exact fare", reminding passengers, to whom it might not have been immediately obvious, that they were about to board a one-man operated bus. *London Transport Museum*

Interior views showing the general similarity to CR 1 at the back end of the vehicle but the very different layout at the front where the doorway is now closer to the driver, though still not conveniently enough placed to prevent him having to swing round in his seat to take fares and issue tickets. The general similarity in internal décor to the RT is very evident although this does not extend to the seat upholstery which is still of the plainish green styling used on STLs. Taken just before the vehicle's entry into service, these views show the lamp shields fitted to comply with wartime blackout regulations. *London Transport Museum*

Inside the vehicle all 20 seats now faced forwards and the original design was retained whereby the rear 12 seats were set at a raised level. Passengers were treated to a new décor which provided a yellow top half to the curved window shrouds surmounted by a similarly coloured wooden moulding with a thin inset red band, a styling which had recently been introduced on prototype RT 1 and which was to become almost a trademark of London Transport design in years to come on the huge post-war RT family of double deckers. Body numbers 224-271 were allocated to the new vehicles, though not applied strictly in sequence, and somewhere along the line a quarter of a ton had been shed in weight compared with CR 1, leaving the production batch at 4tons 5cwt.

The first of the new CR fleet to enter service were CRs 2 and 3 which were licensed at Kingston on 9th September 1939 with all the now-statutory lamp masks and white markings in place. The roofs and back domes, which had been painted black when new, were now repainted grey and only the black edging to the back windows remained as a reminder of how they would have looked in better times History was made with this new batch of CRs which represented the first production run of rear-engined buses ever to be built and operated in Britain. London Transport was now truly at the forefront of passenger vehicle design. The prototype CR 1 may not have been the first rear-engine bus in service – this honour had gone to Midland Red with three vehicles in 1935/36 – but theirs had been purely experimental and no operator had previously taken the gamble of committing themselves to ordering a production run of such vehicles.

The fleet at Kingston gradually built up to eleven vehicles as new ones became available, whereafter further garages were equipped, with batches starting work on 1st November 1939 at Windsor, Hounslow and Uxbridge, and completed by a single vehicle at Enfield on 1st December. The original allocations were as follows:

Kingston (K)	CR 2-11, 19
Windsor (WR)	CR 12-17
Hounslow (AV)	CR 18, 20-27
Uxbridge (UX)	CR 28-34
Enfield (E)	CR 35

The vehicles at Windsor were, of course, in green livery, and by a series of consequential moves they brought about the withdrawal of the last BD class Bedfords. The Central Bus CRs all replaced Dennis Darts, either directly or through transferring Cubs around, and the last operational Dart, DA 10, was withdrawn at Enfield on 1st December. The single CR at Enfield was soon joined by another, CR 11 ex-Kingston, on 6th December. Construction of the remaining CRs continued through to the end of 1939 with the last of all, CR 49, being taken into stock on the first day of 1940. With no work immediately available for the later deliveries there was no option other than to place them in store until such time as they might be required.

Gradually work was found for several of these surplus CRs. Six of them were sent at various dates to Kingston, Uxbridge and Hounslow to augment their existing fleets by providing additional breakdown cover. CR 49 was sent to operate from Harrow Weald on 6th February 1940, but CR usage there lasted only until July of the same year. On 18th February 1941 previously unused CRs 38 and 41 were allocated for use by the Paymaster at Chiswick tram depot, being officially rehoused at Mortlake from 5th June 1941. In addition to these two, CR 6, formerly in service at Kingston, was also transferred to the Paymaster on the same date but was subsequently relicensed as a lorry with all except two seats removed for the use of the Stores Department from 1st May 1941 through to 1st June 1942. CR 42, minus its seats, became a lorry for London Aircraft Production at Chiswick from 1st April until 1st May 1941 but saw no further wartime use. CR 36, 40, 45 were licensed for unspecified work at Chiswick in 1940 and 1941, leaving CR 44 as the only one of the class never to be used for any purpose at all until after the war.

From a passenger viewpoint the CRs were pleasing little vehicles to ride on for much of the time, purring along more quietly than conventional front-engined Cubs which probably says a lot for the quality of the sound insulation within the engine compartment. Every so often, though, this pleasant progress would be disturbed by a violent – and sometimes very noisy – jerk associated with gear changing.

Passengers may have found the CRs pleasant enough, but from an engineering standpoint their introduction into service proved to be by no means a happy experience. Almost from the start, breakdowns and loss of service mileage began to occur, and from as early as 13th September 1939 CRs began going back to Leyland's works at Kingston for adjustments and repairs to be carried out under warranty on a whole variety of items – clutches, gearboxes, brakes, but mainly engines. Incredible as it may seem now, a regular stream of returns to Kingston built up which lasted for a full year, right through to 5th October 1940. Most CRs returned to Kingston on multiple occasions, the record being held by CRs 2 and 4 which both

Photographed loading up at Waltham Cross on 1st December 1939, the sole CR allocated to Enfield garage from new was CR 35, from where it ran alongside conventional Cubs for the next nine months before being delicensed for the duration of the war. Like so many others of these sturdy but unreliable vehicles it was destined to do comparatively little work, amassing a mere 2 years 4½ months of passenger service during its time in London. *D Evans © Omnibus Society*

CR 2 was one of the first two production batch CRs to enter service, on 9th September 1939. The general similarity of the rear end of these vehicles to CR 1 was immediately apparent and the polished strips at the back covering the radiator mesh were a nice design touch. This photograph, taken at Sunbury station, shows the remnants of the original black roof around the back window frames which were left when the remainder was painted in wartime grey prior to entering service. *Jack Taylor*

returned there no fewer than seven times, the total number of referrals during the 13-month period being 135. The only two never to be returned to the manufacturer were CRs 46 and 48, and this was probably because neither of them was used within the warranty period.

In addition to the referrals back to the Kingston factory, CR 3 was sent to Leyland's main works in Lancashire from 1st December 1939 to 15th January 1940, followed by CR 18 between 24th June and 10th August 1940. Presumably work of a more investigative nature was carried out there to identify long-term solutions to some of the problems. A consequence of CR 18's trip to Lancashire was that, upon its return to London, it was sent for storage at Bull Yard where it was among the large number of buses and coaches destroyed by bombing on the night of Saturday 17th September 1940, so the benefits of its time at Leyland, whatever these might have been, were never put to the test.

Despite Leyland's best efforts, engine problems such as crankshaft bearing failures, and side effects from excessive oil consumption, continued to be experienced and eventually a fundamental design problem was found to be at the root of much of the trouble. The power unit, including engine, gearbox and rear axle, was all bolted together and suspended in the frame at three points which failed to take into account the fact that the rear axle torque reaction, due to inertia loading, was particularly heavy at the moment the clutch engaged. As a result, the whole unit tended to pivot about the worm wheel centre line and violent vibration was set up when the clutch was engaged, causing breakage of the chassis cross members and other damage as well as the violent shaking that passengers sometimes objected to. An initial scheme for remounting the engine unit to avoid these problems was devised and applied in about May 1940 to CR 31 based at Hounslow garage, but this only provided a temporary solution as a different type of vibration then developed which became so intense that, under certain circumstances, the 2nd gear was shaken out of mesh. While London Transport was still experimenting to solve this latest setback Leyland came up with a simpler solution involving stop rubbers fitted to the engine supports nearest to the rear axle which at least helped to mitigate this problem.

Salvation from the CR troubles came in 1942. The ongoing programme of replacing 20-seaters with larger buses wherever passenger demand required and road conditions made it possible had now brought London Transport to the position whereby it no longer needed to use any of the CRs, and arrangements were made to withdraw them all from service, with a decision on their long term future deferred until after the war had ended. The two at Enfield garage had been delicensed as early as 1st September 1940 and all the green ones at Windsor had ceased operation by 1st May 1941. After the start of 1942 just 33 CRs were still licensed for passenger use and most of these were taken out of service in batches on 1st May, 1st June and the final three (CR 9 at Kingston and CR 37, 49 at Hounslow) on 1st July. Many were sent straight for storage in the old AEC works at Forest Road, Walthamstow, and others which had been temporarily housed in surplus space at Alperton, Leyton, Old Kent Road and West Green garages later joined them there so that by 1944 33 CRs had been gathered into this one location. Of the remainder, six spent the rest of the war years in Mortlake garage and three at Staines, while the six green ones stayed where they had last operated, at Windsor.

Sheer desperation brought the CRs back into service in 1946. London Transport's total inability to find sufficient serviceable buses to satisfy the early post-war boom in passenger traffic led to a decision to employ 20-seaters – the only class of vehicle in which it had a surplus – as extras on crew-worked Central Bus peak hour services. As recorded in an earlier chapter, some C class Cubs were involved but CRs inevitably formed the majority. In October 1945 the CRs stored in Mortlake were sent to join the majority in the Forest Road storage yard followed by those at Staines and Windsor in March and April 1946. There, a team of staff worked to

Wearing Lincoln green and white colours received at its first and only overhaul in December 1939, CR 1's re-entry into service at Streatham on 15th June 1946 came as a great surprise since most other prototypes and general oddities were disposed of as soon as possible after the war ended. Photographed at the Liverpool Street terminus of route 133 outside Broad Street station, the fact that CR 1 still displays its white wartime markings well after the war's end suggests an air of semi-neglect, while the ill-fitting engine compartment door does little to inspire confidence.
G B P Martin

Another Streatham-based vehicle, CR 17, looks in dire need of a fresh coat of paint, at least on its roof where the top coat of brown, hurriedly slapped over the earlier covering of grey, has clearly seen better days. It was not until December 1948 that CR 17 finally went in for overhaul and it subsequently became one of the fortunate members of the class lucky enough to stay in service until as late as 1953. *W J Haynes*

bring them back into serviceable condition. After years of disuse many wore a very dilapidated appearance and some still retained grey roofs from the early wartime years, though most roofs had been repainted brown during wartime overhauls. An exception to the general air of scruffiness was presented by CRs 6, 11, 19, 23, 25, 26, 29, 41 and 48 which all returned to service newly repainted and looking quite resplendent. These nine vehicles had been affected by a fire which broke out in the premises of the Ever Ready battery company next door to the Forest Road yard on Monday 6th May, and though not seriously damaged they could not be returned to service until after smoke damage had been made good.

The only CR which did not find itself at the Forest Road yard at this time was green CR 14 which was loaned to the CAV organisation at Acton from 30th May through to 13th November 1946, the purpose of which is not known. When it subsequently re-entered service (at Camberwell garage) in August 1947, much later than all the others, it was in red livery and carrying the body previously on CR 6, the only instance of body-swapping ever to occur with the CR class. For the majority of CRs the day of reawakening came on Monday 15th June 1946 when 37 of them were licensed for service at eight garages: Athol Street (5), Chalk Farm (4), Cricklewood (4), Croydon (4), Merton (4), Putney Bridge (4), Streatham (6) and Tottenham (6). The six at Streatham stood out from all the others in being green and, surprisingly, the prototype CR 1 was included amongst their number until 1st December 1947 when it was withdrawn from service, being subsequently dismantled in February 1949. Several more came into service as and when they were ready, but the position did not remain static. Athol Street lost its CRs as early as July 1946 and two further garages (Barking and Forest Gate) joined the ranks of regular CR users, while Elmers End, Leyton and Hornchurch used them for short periods in 1946, and Dalston in 1947.

The reappearance of the CRs on a host of Central area routes was well publicised in the newspapers of the time, but this did not stop them looking totally incongruous working alongside six-wheeled double deckers and the like, and when fully laden with passengers, as they often were, they were reminiscent of strange little busy beetles scurrying about their daily business. Once on board, it seemed most odd to see conductors walking around collecting fares on such tiny little vehicles and journeys on them were often quite an adventure, being punctuated by raucous gear-changes often accompanied by a series of neck-crunching jerks. Passengers remained blissfully unaware that, back in the garages, it was proving something of a nightmare to keep the little CRs on the road, to the extent that many of them spent far too much time tucked away in odd corners engineless and waiting for a hapless engineer to find time to try and sort out their problems.

Towards the end of 1947 CAV asked to borrow a CR for a second time for development and testing purposes. This time CR 28 – which had been in service at Tottenham and Dalston for about six months – was selected. It went to the Acton factory on 10th November 1947 and was there for much longer than its predecessor, not returning to Chiswick until 8th March 1950. In a fairly dilapidated condition by now, it never returned to service and was broken up in November 1951.

As the need for CRs on Central Bus services diminished during 1948 another call for their services arose in the form of the Olympic Games which were held at Wembley and required a fleet of buses to be on hand to convey competitors and staff. Up to 30 CRs were made available alongside a miscellany of other vehicles between 29th July and 14th August 1948, and these were divided between Chelverton Road, Cricklewood, Hounslow and Putney Bridge garages. Following on from this,

In contrast to some of the class, CR 42 operating from Forest Gate garage, looked very smart when photographed at Victoria bus station in March 1949 in the company of a post-war STD and a couple of STLs. It had emerged from overhaul as recently as 14th December 1948 and certainly outshone in appearance the STL standing immediately behind it. Having run in passenger service only between 15th June and 30th September 1946 and 15th December 1948 to 12th June 1949 inclusive, CR 42 recorded the shortest working life of all at a mere 9½ months. *Allen T Smith © A B Cross*

Just occasionally the Central Bus department's CRs took a post-war turn at the one-man operation for which they were originally intended on the last two remaining omo routes, 238 and 252. On this occasion a sad and rather battered-looking CR 30 is found at Gidea Park station on route 238. The date is unrecorded but it would have been at some time between October 1948 and May 1949 while CR 30 was based at Hornchurch garage. *Alan Nightingale collection*

A surprise return of CRs to the Country area began with a pair which started work at East Grinstead on 3rd October 1948. One of these, CR 14, is seen at Oxted station in the green livery it received on overhaul in February 1949 having temporarily been red since receiving the body from CR 6 in August 1947. This had been the only instance of CRs ever exchanging bodies. *J H Aston*

odd ones appeared at suburban garages such as Hornchurch, Kingston, Sutton and Uxbridge during the latter part of 1948 and early 1949 in addition to a reduced number now used as Central area extras. CRs 11 and 30, which were at Hornchurch during this time, could sometimes be found on this garage's routes 238 and 252, temporarily reviving one-man operation of CRs on Central Bus work for the first time since 1942.

Starting on 1st May 1949 a concerted effort was made to remove all CRs from the Central Bus fleet: 24 were withdrawn during May, 9 in June and the final 6 on 20th July 1949. For the record these last six were CR 41 at Chalk Farm, CR 11 at Hornchurch, CR 33 at Merton, CR 9 at Putney Bridge, CR 10 at Streatham and CR 17 at Tottenham. Upon delicensing, 22 were despatched to the closed garage at Slough to join three others already stored there and the remaining 17 were laid-up at Reigate.

At the same time as the Central Bus department was planning to dispense with its CRs as soon as possible, its Country Bus counterpart was in the process of offering them a new, if limited, lease of life augmenting existing one-man services or commencing new ones. This process started when red CR 14 and green CR 16 were sent to East Grinstead garage for route 494 on 3rd October 1948, and when CR 14 emerged from its post-war overhaul at the start of February 1949 it, too, had received green livery. Meanwhile CRs had also become established at Chelsham on routes 464, 465 and 485 with the arrival of red CR 5 on 24th November 1948. CR 13, now newly repainted green on overhaul, replaced it at the beginning of February 1949 and in March CR 19, now also green, joined it at Chelsham. This new era of CR operation, which saw them running as one-man buses on a regular basis as originally intended for the first time since 1942, suddenly came to a halt when all four were withdrawn and put in store at Reigate. This occurred on 20th July 1949, the same date that Central Bus operation of CRs had ceased.

In the background, within the higher echelons of the London Transport organisation, much discussion had been going on about what to do with the CRs. B H Harbour, Operating Manager of the Country Bus & Coach department, still wished to retain a few to meet his department's possible needs until new vehicles could be obtained, but opinions about what should be done with the remainder varied. The only area of agreement was that if surplus 20-seaters were to be disposed of these should be CRs and not front-engined Cubs. An invitation had already been sent to other BTC companies to tender for surplus CRs but this had evoked no interest whatever, although a few of them expressed the worry that if the CRs were disposed of to outside operators they could be used in competition with state-owned companies. Next, a wider invitation was issued for expressions of interest and circulated to no fewer than 162 outside operators and dealers, but this resulted in only two enquiries; these were from Lammas Motors (a dealer) offering £65 per vehicle and Derby Corporation who stated that they might be interested in obtaining two at £150 each.

On the very day that the last CRs were withdrawn from service, 20th July 1949, a meeting of London Transport's road services officers minuted: "Noted the desirability of disposing of later CR-type vehicles in preference to earlier vehicles of normal forward engine design was occasioned largely by operational difficulties since (among other reasons) the driver had frequently to leave his seat to collect fares in the former". A suggestion that the CRs were suitable to be retained for private hire and school journey contract work by one officer was flatly rejected by another, and a memorandum to the Executive on 5th October 1949 that 42 out of the 47 CRs should be sold was turned down by the Executive eight days later who decreed that they should not be disposed of but must remain in store until further notice.

The sharp slope into Reigate garage finds CRs 31, 12 and 38 awaiting their fate in October 1951 and prompting the observation that storage at this angle for any length of time is unlikely to do their springs much good! They had, in fact, been brought to Reigate two months earlier having previously been in store at Slough since 1949, whereupon they were prepared for their sale to L W Vass in November 1951 and a new life in Ceylon. *Ken Blacker*

Two of the CRs featured in the previous photograph are present in the South Western Omnibus Company's yard at Ratmalana in Ceylon on 15th August 1955. CR 38 is the one nearest the camera with CR 31 next to it; the third CR is unidentified. The CRs were not totally successful in Ceylon and only two out of eight still remained in the South Western fleet at the time the island's bus services were nationalised at the start of 1958. *British Commercial Vehicle Museum*

Vass's yard in Bedfordshire was seething with CRs in late 1951, just a few of which can be seen here. Notices in the bulkhead windows state "Supplied by L W Vass Ltd, Ampthill Bedford, Eng". All signs of identity have been removed but the smartly painted one in the centre carries a blind reading "Aryian, Job, Port Harcourt" indicating that the intended buyer was in the British West African colony of Nigeria. Beyond that is one with the blind showing "Commercial Vehicles Auckland NZ" which may indicate that it was CR 47, the only one known for certain to have gone to New Zealand. *Alan Nightingale collection*

The matter was raised again in May 1950 when it was asserted that the position regarding spare parts for maintenance of the CRs had deteriorated further and, so far as the manufacturer was concerned, these vehicles were obsolete. This time approval to dispose of the 47 vehicles was forthcoming, subject to the sale being restricted to 37 "for the time being".

Disposal of the surplus CRs finally began in July 1951, an offer having been received from a Bedfordshire dealer, L W Vass of Ampthill, for 36 of them at a good price of £250 apiece, for future export to an unspecified destination in East Africa. In fact, Vass took just 24 between July 1951 and May 1952 while another three were broken up at Chiswick. Vass's anticipated East Africa venture must have fallen through as many of them remained in the Ampthill yard for many years, although eight passed to the South Western Omnibus Company of Ceylon in February 1952 while others finished up with the large Lefkaritis organisation in Larnaca, Cyprus, and one was traced to Auckland, New Zealand.

Great excitement stirred in the enthusiast world, and was replicated in sections of the transport press, in September 1951 when, after lying moribund for two years, a few CRs were put back into passenger service. After seeing numerous specimens of the class looking semi-derelict at Reigate garage, where most of them were prepared for sale, the concept of witnessing a seemingly defunct class being coaxed back into life seemed too good to be true. Quite unexpectedly, the opportunity had arisen once again to glide along in a CR, its smoothness punctuated as ever by the odd violent gearchange, perched perhaps on the back 'shelf' which was always a favourite warm seat for children but quite often for adults, too.

The second and final CR revival began on 26th September 1951 and CR 4, having done little work since its last overhaul in December 1948, still presented quite a smart appearance albeit in red livery. Allocated to Epping garage, it could be found on routes 381 and 393 and is seen on this occasion with RT 1024, which had been new to Epping in November 1948, standing immediately behind.
J H Aston

This mini-renaissance began on 26th September 1951 when red CRs 4 and 43 replaced normal-control Cubs at Epping on routes 381 and 393, and green CRs 13 and 14 reintroduced the class to route 494 at East Grinstead. A few days later, on 1st October, another bus was found necessary at Epping and this was CR 10 which had been repainted green as recently as April 1949. The Epping operation lasted less than a year with CRs 10 and 43 departing on 1st June 1952 followed by CR 4 on 1st September. CR 10's destination marked yet another new sphere of operation for the class, at Tring garage on route 387 where it was joined on the same day by relicensed CR 17.

With new GSs coming off the production line, 1953 marked the final year of operation for this now much depleted class, but it did not pass without note. As a kind of swansong to a dying breed, yet another sphere of operation was commenced when CR 43, now newly repainted green, inaugurated new Epsom local service 481 operated by Leatherhead garage on 6th May 1953. One vehicle quickly proved to be insufficient and CR 13, previously at East Grinstead, joined it on 22nd June. CR 4 was relicensed at Tring on 6th July replacing CR 17 whose road tax expired on 1st August, leaving five CRs in service up to the very end. The last day they were available for service was Saturday 31st October 1953, although they may not have all run on that day. Conventional Cubs replaced CR 4 (the only red one still in use) and CR 10 at Tring along with CR 14 at East Grinstead on 1st November, while CRs 13 and 43 at Leatherhead gained the doubtful honour of being the only ones of their type ever to be directly replaced by new GSs.

Subsequent to the 1951 disposals, 20 CRs had remained in stock, sales of which to dealers recommenced in January 1953 and were completed with the departure of CRs 6 and 10 to North's of Leeds on 6th July 1955 (although these two were actually broken up on a travellers' site just off Sevenoaks Way in St Mary Cray, Kent and never actually reached North's premises in Leeds, a fate which also befell STDs 94, 95 and 99 as well as several dozen STLs). However, this dealer was successful in selling other CRs to the Lefkaritis fleet in Cyprus. An exception to the sales to dealers was CR 14 which was officially transferred to museum stock on 25th March 1954 and sent for storage at Reigate along with other museum vehicles.

Even in their final years of existence the CRs were capable of springing the odd surprises such as on this occasion, on 3rd March 1953, when Guildford garage covered a temporary Cub shortage by borrowing CR 14 from East Grinstead. This may well have been the first and only time a CR ever ran from Guildford, and it would have been interesting to know what the driver thought of his unusual steed. *Alan B Cross*

With the end now so close, it was amazing to find CR 43 being used for a new service start-up on 6th May 1953 in the form of Leatherhead-worked Epsom local 481. Furthermore CR 43 had been newly repainted for the occasion, probably at Reigate, and became by far the highest numbered CR ever to appear in green livery. Unlike all other buses repainted since the end of August 1950 it was not in the rather drab 'all green' but reflected traditional styling of earlier days. *Peter Mitchell*

Viewed in retrospect, the CRs probably marked one of London Transport's biggest rolling stock blunders. Like the superb TFs, to which they bore a strong frontal resemblance, they came from the Leyland stable, but whereas the TF nudged gently and successfully at the boundaries of technological development, the CR fell short in so many ways. It was almost as though the design teams at Chiswick and Leyland, excellent as they were, were determined to push technology forward simply for its own sake without pausing to consider the cost or practicalities of what they were doing. The design of the CR, particularly in relation to the engine and drive arrangements, was far too complicated and could hardly be justified on a 20-seater where simplicity should have been the keynote, and London Transport's engineers were guilty of a crass omission in driving ahead with production without first taking the simple precaution of ensuring that the initial prototype worked properly. In addition to all its technical deficiencies, the CR was not even really suited to operation as a one-man operated vehicle because of its internal layout. Despite all the money spent on designing and manufacturing the CRs the stark fact emerges that vehicles in the production batch CR 2-49 only ran as one-man buses for an average of 1 year 11¾ months during their entire time with London Transport. In fact, seven of them – CRs 36, 38, 40-42, 44, 45 – were never used as one-man buses at all. The doubtful accolade for spending the least time in any passenger carrying capacity went to CR 42 which ran in service for merely 9½ months.

We are extremely lucky that two of these strange little buses remain to this day in preservation. For many years it was thought that CR 14 would be the only one to survive after London Transport so wisely set it aside for posterity in 1954. In 1960 it passed, still in state ownership, to the Museum of British Transport at Clapham and, rather unusually, it was sold to private preservationists Tim Nicholson and Prince Marshall in 1967. It has since returned to the modern equivalent of its ancestral home and is now in the London Transport Museum reserve collection at Acton. In 1979 Group Captain Brian Lemon, who was serving in Cyprus, located with the help of John Durey the semi-derelict remains of CR 16 lying dumped in a yard belonging to the Lefkaritis company. It was one of the vehicles that had been exported to the island by L W Vass in 1952. Repatriated to the UK, it spent many years with Malcolm Shevington undergoing restoration, firstly into red livery but nowadays to its original green. It is now owned by the London Bus Preservation Trust and can be viewed in the London Bus Museum at Brooklands.

The end came for the CRs on 31st October 1953 and a week later green CRs 10 and 17 are in Tring garage yard awaiting their fate. After a final journey to W North of Leeds both probably ended up being scrapped. CR 10 had a claim to fame in having notched up the most time in passenger service of all CRs. A grand total of 7 years 1 month 3 weeks in service equated to almost exactly half the time it had spent in the fleet up to the time of its final withdrawal. *John C Gillham*

Newly emerged from its lengthy restoration, CR 16 presents enthusiasts with an opportunity to sample one of these unusual little vehicles in action. The photograph was taken outside the old London Bus Museum premises at Cobham showing CR 16 carrying the green livery that it always wore for service in London. It is now on permanent display at the London Bus Museum at Brooklands. *Barry LeJeune*